John Nisbet

Studies in Forestry

Being a Short Course of Lectures on the Principles of Sylvi-Culture Delivered at the

Botanic Garden, Oxford

John Nisbet

Studies in Forestry
Being a Short Course of Lectures on the Principles of Sylvi-Culture Delivered at the Botanic Garden, Oxford

ISBN/EAN: 9783337073640

Printed in Europe, USA, Canada, Australia, Japan

Cover: Foto ©berggeist007 / pixelio.de

More available books at **www.hansebooks.com**

STUDIES IN FORESTRY

BEING

A SHORT COURSE OF LECTURES

ON THE

PRINCIPLES OF SYLVICULTURE

DELIVERED AT

THE BOTANIC GARDEN, OXFORD

During the Hilary and Michaelmas Terms, 1893

BY

JOHN NISBET, D.Oec.

OF THE INDIAN FOREST SERVICE

AUTHOR OF 'BRITISH FOREST TREES'
EDITOR OF THE SIXTH EDITION OF BROWN'S 'THE FORESTER'
TRANSLATOR OF KAUSCHINGER AND FÜRST'S 'PROTECTION OF WOODLANDS,'
AND OF HARTIG'S 'ANATOMY AND PHYSIOLOGY OF PLANTS,' ETC.

Oxford

AT THE CLARENDON PRESS

1894

PREFACE

THE various chapters of the present work formed two short courses of lectures on the Principles of Sylviculture, which were delivered, during the Lent and Michaelmas Terms of last year (1893), at the Botanic Garden in Oxford, where Professor Vines, F.R.S., was good enough to arrange for their delivery in his class-room. I am further indebted to him for kind assistance given during the revision of the proofs, and gladly take this opportunity of tendering my thanks to him on both accounts.

With the permission of the Secretary of State for India, I have been enabled to make several of the chapters more complete than they might otherwise have been, by utilizing freely the greater portion of a series of six *Essays on Sylvicultural Subjects*, which were written by me in Bavaria during 1892, and were published in 1893 by the Government of India for distribution among their Forest Officers.

As will be apparent throughout every chapter, my convictions regarding economic Forestry, i. e. Sylviculture, have been formed in a Teutonic school. **But Science is Truth**; and in acquiring information with regard to the growth of Forest Trees, or concerning any other branch of natural knowledge, it ought to be a matter of perfect indifference from what well this may be drawn. I have therefore had no hesitation in boldly acknowledging the German sources from which many of the lessons I am trying to teach have been learned.

With regard to Forestry, Britain stands in a very peculiar position. Although she has now only a very insignificant area under woodlands in comparison with any other great European power, yet all her national and commercial prosperity is mainly due to her forest wealth—first of all, in the time of the formation of the coal-measures, and later on again in the days of the 'wooden walls of England.' For what could the 'Hearts of Oak' have achieved in the way of naval supremacy without the ships of home-grown Oak? It is not because Germany has a forest area fully eleven times as large as the woodlands and nurseries of Great Britain and Ireland that we must look mainly to that country for instruction and guidance with regard to Forestry; it is because at Universities, Academies, Institutes, and Forest Schools, many Professors of high scientific attainments are engaged not only in teaching the Science of Forestry and the Art of Sylviculture, but also in individual and conjoint efforts and investigations for ascertaining the natural laws which regulate the growth of timber-crops.

In Britain, however, nothing of this sort is going on. There is no body of men of anything like good scientific attainments who are known to be carrying on investigations that have for their object the improvement of our natural knowledge regarding the growth of timber. Times have changed since one of the oldest and most remarkable of the English works on Forestry, John Evelyn's classic *Silva or a Discourse of Forest Trees*, was read by him in its first form as a paper before the Royal Society in October, 1664; for at that time we were certainly abreast of continental knowledge. But, with the subsequent decrease in the woodland area throughout Britain, we have lagged far behind continental countries with regard to the improvement of our knowledge of Forestry. With the present shrinkage in the agricultural and pastural value of land, however, there certainly is now scope for applying a knowledge of economic and purely financial Forestry, i.e. Sylviculture, to considerable tracts that are now lying waste and unproductive.

It may perhaps be convenient, for the sake of ready reference, to give a short list of the chief works that have been frequently quoted throughout the various chapters of this book. These are :—

GREBE, *Die Betriebs- und Ertragsregelung der Forste*, 2nd edit., 1879.
NEY, *Die Lehre vom Waldbau*, 1885.
WEBER, *Die Aufgaben der Forstwirthschaft*, in Lorey's *Handbuch der Forstwissenschaft*, vol. i. 1886.
WILLKOMM, *Die Forstliche Flora von Deutschland und Oesterreich*, 2nd edit., 1887.
HESS, *Der Forstschutz*, 2nd edit., vol. i. 1887 ; vol. ii. 1890.
GAYER, *Die Forstbenutzung*, 7th edit., 1888.
—- *Der Waldbau*, 3rd edit., 1889.
R. HARTIG, *Lehrbuch der Baumkrankheiten*, 2nd edit. 1889.
—— *Anatomie und Physiologie der Pflanzen*, 1891.
C. HEYER, *Der Waldbau*, 4th edit., vol. i. 1891 ; vol. ii. 1893.
RAMANN, *Forstliche Bodenkunde und Klimatologie*, 1893.

J. NISBET.

145 NORWICH ROAD, IPSWICH :
    March, 1894.

CONTENTS

Contents

CHAPTER IX.

CHAPTER X.

CHAPTER XI.

CHAPTER XII.

CHAPTER XIII.

CHAPTER XIV.

STUDIES IN FORESTRY

CHAPTER I

FORESTRY IN BRITAIN

In the *Agricultural Returns of Great Britain*, published for 1892 by the Board of Agriculture, it is shown that, whilst the total area of the United Kingdom is 77,642,099 acres, the extent to which it is utilized agriculturally and sylviculturally is as follows :—

Total Area of Land and Water	77,642,099 acres	Percentage of Total Area
Arable Land	20,444,577 ,,	25·5
Permanent Pasture . . .	27,533,326 ,,	35·5
Woodlands (and Nurseries) .	3,005,670 ,,	3·8

Though the acreage covered with forest is undoubtedly extensive, amounting as it does to 470 square miles, or 3·8 % of the total area of land and water, yet it does not at first glance convey the full extent of the important interests represented by it. On arable land, and in fact on all land utilized agriculturally, the sowing of the seed is followed within one year by the reaping of the harvest; but in woodlands such can never be the case. With the exception of willow-beds cut over annually when once the stools are in vigorous bearing, and coppice-hags of Oak for tanning-bark, or of various woods for other purposes, long periods must necessarily elapse before the timber-crops reach their highest material, technical, and financial value; hence the advantageous time of maturity may be said roughly to vary from about 60 to 80 years with conifers, up to about 150 years for Oak. But taking a rotation of about 90 years as the average,—and that would certainly not.be too high a one,—this would mean that the present market value of the timber-crops on these three millions of acres of woods (assuming that they range normally from 0 to 90 years of age) is equal .to 3,000,000, multiplied by the average quantity of timber contained in crops 45 years of age, or equal to 1,500,000 acres multiplied by the average quantity of timber yielded by a mature crop of 90 years in age.

Without entering into detailed figures, which of course vary essentially according as the crops may be of Scots Pine, Spruce, Silver Fir, Oak, Beech, &c., or mixed woods, it will be at once intelligible that the monetary value represented by the timber standing on our three million acres of woodland is of sufficient amount to be decidedly of national importance.

At the very lowest computation, *the actual cost of production of the timber-crops*, which, even under management of an unsatisfactory economic nature only, must be far below their true market value, equals

((annual rental of 3,000,000 acres)

$$+ \text{costs of forming or regenerating 3,000,000 acres}) \; 1 \cdot 0 \; r^{\frac{0 \cdot 0}{2}}$$

where $r=$ the rate per cent. at which the proprietor is content to lock up his capital represented by soil plus growing-stock. If the sylvicultural soil were of equal productive power with that utilized agriculturally, then it would be necessary to estimate r at a higher rate than is permissible in actuarial calculations relating to farming, owing to the greater risk run by forest crops from wind, fire, insects, &c., before attaining their financial maturity. Thus, in central Germany[1], whilst the rate of interest for ordinary investments is about $3\frac{1}{2}$ to $4\frac{1}{2}$ %, that for actuarial calculations referring to agricultural soil ranges from

---

[1] See G. Heyer, *Anleitung zur Waldwerthrechnung*, 4th edit. 1892' p. 15.

2 to 3 %, and for State Forest land it may be put at 2 to $2\frac{1}{2}$ % ; but rates varying from $2\frac{1}{2}$ to $3\frac{1}{2}$ % are laid down in laws referring to expropriation of forest areas, whilst Pressler[1] insists that it should be $3\frac{1}{2}$ % for State Forests, 4 % for large land-owners, and 4 to 5 % for smaller proprietors working on more purely financial principles. Owing, however, to the extrinsic appreciation in the value of broad acres, as conferring a certain degree of social elevation among the moneyed classes in Britain, so low a rate as $2\frac{1}{2}$ % may perhaps be taken for the purpose of determining the actual minimum cost of production of our growing-stock of timber[2]. Taking the annual average rental of the land at about 5*s.* an acre, and the cost of forming or regenerating the crop at £2 an acre (which was the sum named by Mr. Macgregor in his evidence before the Committee on Forestry on June 28, 1887, as the cost of planting Scots Pine in the Athole forests in Perthshire), the actual cost of production of our woodlands at the present time will therefore be

$$(3,005,670 \times £2\ 5s.)\ 1 \cdot 025^{45} = £20,544,571.$$

This does not take into consideration intermediate returns from thinning, &c. ; but, *per contra*, it omits all annual charges for administration, protection, and rates chargeable. And this is only the *actual cost of production*, whilst the *prospective value of the mature crops* is in reality a very much greater sum. Thus the forests of Germany aggregate 34,353,743 acres[3], in regard to which Professor Gayer, in his rectorial address before the University of Munich in 1889, stated[4] that their annual outturn in timber amounted to about 60,000,000 cubic métres worth from £20,000,000 to £22,500,000 sterling, ' *so that, reckoning 2 % as the rate of interest yielded, the capital value of*

---

[1] *Rationeller Waldwirth*, vol. i. 1858, p. 10.

[2] According to Prof. Endres the present returns from forestry in Germany lie between 2 and 3 %. (Conrad, Elster, Lexis, and Loening's *Handwörter-buch der Staatswissenschaften*, 1892, vol. iii. p. 602.)

[3] *Agricultural Returns for* 1892 (Board of Agriculture), p. 167.

[4] *Der Wald im Wechsel der Zeiten*, 1889, p. 15.

*all the German forests may be assessed at about £1,000,000,000
sterling.'*

Professor Weber also states [1] that, in addition to £4,150,000
annually spent in the management, reproduction, and protection of the forests, and in the felling, preparing, and handling of
the produce before it is delivered into the hands of the buyer,
the timber and other produce of the woodlands directly affords
employment to 583,000 persons, or 9 % of all the industrial
classes throughout the empire, who are engaged in industries
dependent on the forests for the raw material requisite for
their various trades. These 583,000 bread-winners probably
represent about 3,000,000 of the population. And in addition
thereto, there must also be taken into account the enormous
sums whose outlay is necessitated for transport by land and water
after the raw produce has come into the hands of the buyer.

It therefore follows that, if our woodlands were as economically and as well managed as the German forests, they would
annually yield very nearly £2,000,000, and have a capital value
of £90,000,000, adopting 2 % as the rate of interest yielded, or
at any rate about £50,000,000, even adopting only twenty-five
years' purchase as their value, and presuming that they yielded
as much as 4 % per annum on their capital value of soil plus
growing-stock.

Our woodland tracts are by no means equally distributed
over the length and breadth of the land ; for, as Major Craigie
reports in the Board of Agriculture's *Returns* for 1891
(page x.) :—

'The county of Hants now stands in the Returns as possessing the
largest area of woodland in England, or 122,574 acres. Sussex, with
122,073. comes second, while *it may be worth noting that the four counties
of Hants, Sussex, Surrey, and Kent, forming the south-eastern corner of
England, possess between them nearly a fourth of the English woods and
plantations, showing over 11 % of their surface thus occupied,* in contrast
with 4 % as the mean of the rest of the country.

[1] *Die Aufgaben der Forstwirthschaft* in Lorey's *Handbuch der Forst-wissenschaft*, 1886, vol. i. p. 85.

' In Scotland the county of Inverness accounts for 169,000 acres of woodland. This area is far the largest in Great Britain. *It is considerably in excess of the surface returned as under all forms of crops or grass in that county, and nearly equal to a fifth part of the whole Scottish woodlands.*'

To look at Scotland alone somewhat more closely, the area under woods and nurseries amounted to 876,250 acres out of a total of 19,453,843 for land and water, or in other words we find—

nearly *one-fourth* as much woodland as arable land (3,550,095 acres) ;

nearly *two-thirds* as much woodland as permanent pasture (1,338,249 acres) ;

more than *two-thirds* as much woodland as land under corn crops (1,297,231 acres) ;

and nearly *forty* % more woodland than arable land under green crops (638,794 acres).

The timber and minor produce produced on this area by home growth is, however, entirely inadequate to supply our actual requirements, as may be seen at a glance from the statement of Imports into and Exports from the United Kingdom during the last three years, compiled from the Customs *Returns* for 1892 [1] (see next page).

Although from these *Returns* it is hardly possible to eliminate all the classes of timber which it would be physically impossible to grow in our climate, as, for example, the Teak timber imported from Burma for the lining of iron ships, and the Australian hardwoods used for street-pavement in London, there have been left out of account the imports of Mahogany (amounting in 1892 to 56,315 tons = £501,203) of Cutch and Gambier (25,192 tons = £548,395), of Caoutchouc and Gutta Percha (317,660 cwts. = £3,501,923), &c., the supplies of which must, under all circumstances, be drawn from across the sea.

It would of course be absurd to throw out any broad casual statement that, if the treatment accorded to our woodlands were such as might be dictated by a better knowledge of the laws governing tree-growth, and of the principles underlying

[1] *Accounts relating to Trade and Navigation of the United Kingdom,* December, 1892.

IMPORTS.	1890.		1891.		1892.	
WOOD AND TIMBER.	Loads of 50 cb. ft.	Value.	Loads of 50 cb. ft.	Value.	Loads of 50 cb. ft.	Value.
Hewn (in the round or square).		£		£		£
From Russia	316,952	534,366	289,069	469,007	329,636	517,599
,, Sweden and Norway	673,305	946,599	694,838	940,149	782,677	1,051,715
,, Germany	287,482	727,352	278,084	627,972	288,614	688,087
,, United States	151,697	615,140	149,032	643,171	165,488	669,104
,, British East Indies	46,937	528,450	39,792	414,853	34,897	372,443
,, British North America	180,066	883,461	151,828	703,604	194,654	919,470
,, Other Countries	621,935	769,186	648,934	710,031	673,174	687,428
Total	2,278,374	5,004,554	2,251,577	4,508,787	2,469,140	4,905,846
Sawn or Split, Planed or Dressed.						
From Russia	1,202,222	2,566,905	1,218,025	2,427,588	1,316,258	2,726,113
,, Sweden and Norway	1,970,361	4,338,860	1,853,641	3,776,075	2,046,167	4,224,391
,, United States	308,424	932,863	308,034	859,355	407,854	1,133,771
,, British North America	1,185,205	2,903,524	890,597	2,013,452	1,212,877	2,811,059
,, Other Countries	112,102	350,069	108,155	303,338	111,153	284,807
Total	4,778,314	11,092,221	4,378,452	9,379,808	5,094,309	11,180,141
Staves of all dimensions	155,995	669,243	130,101	590,543	136,063	593,539
Totals for Timber alone	7,212,683	16,766,018	6,760,130	14,479,138	7,699,512	16,679,526
MINOR PRODUCE.						
Pulp of wood for paper	137,837 tons.	766,841	156,464 tons.	848,986	190,938 tons.	981,025
Rosin	1,627,446 cwts.	376,625	1,811,740 cwts.	400,939	1,681,393 cwts.	384,050
Bark for Tanners and Dyers	579,438 cwts.	263,259	489,846 cwts.	216,815	380,337 cwts.	158,105
Total value of Imports	. . .	18,172,743	. . .	15,945,878	. . .	18,202,706
EXPORTS. WOOD AND TIMBER.	Loads of 50 cb. ft.	Value.	Loads of 50 cb. ft.	Value.	Loads of 50 cb. ft.	Value.
		£		£		£
Sawn or Split, Planed or Dressed	19,745	74,678	17,452	64,675	20,935	72,860
Excess value of Imports over Exports	. . .	18,098,065	. . .	15,881,203	. . .	18,129,84[?]

sylviculture (i. e. when forestry is conducted scientifically), we could supply a great many of these requirements by home-grown timber ; but the argument will be strictly pertinent if it be shown in what respects our woodland owners can, and should, strive to compete with foreign importers. Taking the countries in which identically the same species of trees are grown as may be produced in forests in Britain, there still remain the following imports that may be regarded as utilized by us, and not exported again :—

Imported from Russia, Sweden, Norway, and Germany during 1892.		
	Loads.	Value.
Timber in the Rough	1,400,927	£2,257,401
Converted Timber .	3,362,425	£6,950,504
Total	4,763,352	£9,207,905

It may be unhesitatingly maintained that, if due attention were given to the selection of proper species for given soils and situations, and if the principles relating to the most favourable density of plantations, or sowings, or natural reproductions, and to the operations of tending (clearing, thinning, &c.) were *properly understood and practised* throughout Britain, there would be not the slightest necessity for the insertion (as at present obtains) of any clauses into Government contracts stipulating for the use of foreign wood in preference to home-grown timber. But if woods are allowed to grow up so that a considerable portion of the energy of growth of the individual trees forming the crop is dissipated in branch development, in place of being concentrated in the formation of a clean, smooth, full-wooded bole of high technical quality, then no one need be surprised at every person concerned in its utilization giving a solid preference to foreign timber grown under more rational conditions, and consequently of higher technical value, owing to its comparative freedom from branches and knots.

One economic point of great importance may here be noted,

viz. the steady appreciation of forest produce during the past [1]. And this is bound to become accentuated in the future; for whilst population and the demands for timber, &c., are constantly increasing, the woodland area, and consequently the possibility of satisfying these demands, are constantly decreasing over the whole surface of the globe.

On this point Professor Weber remarks [2], in speaking of the increasing demands for forest produce, and the ever-growing utilization of timber as a raw product for various mechanical and chemical industries, that :—

'A proof of the strength of those influences is yielded in the fact, shown by the statistics of prices for the different dimensions of wood, that *the mean annual increase in the price of timber during the last fifty years has been from* 2 *to* $2\frac{1}{2}$ % throughout Germany.'

This consideration is one that should most certainly not be lost sight of when we are face to face with the fact [3] that in fifty years' time the forests of America will have become exhausted, even if the rate of timber extraction does not increase at all beyond its present dimensions. The natural consequences of this must not only be that America will, long before the close of that period, not have any surplus timber to export for the supply of our requirements, but also that she must be a competitor with us for any surplus timber that may then be available, in any other part of the world, for export beyond the limits of the countries where it is grown.

It is not contended that any considerable portion of the £1,365,075 paid away in 1892 for wood-pulp and rosin could have been retained in exchange for the produce of our own coniferous woods; for the manufacture of cellulose, and the collection of rosin, are usually conducted in localities where

---

[1] See G. Heyer, *Anleitung zur Waldwerthrechnung*, 3rd edit. 1883, p. 9.

[2] *Lehrbuch der Forsteinrichtung*, 1891, p. 37.

[3] See Article on American Forests in the *Journal of the Society of Arts* for February, 1893.

wood and labour are comparatively very cheap. But surely whatever oak-bark is required by tanners might quite easily be grown at home on vacant land along the railway lines in the vicinity of the centres consuming it, so as to undersell foreign imports which incur costs of repeated handling and transport before reaching us. New sources of employment might thus easily be created and developed for many thousands of people, especially during the early spring, and partly also in the winter months.

To confine this statement to the practically useful, there is little doubt that if any encouragement were given to the dissemination of sound knowledge concerning forestry throughout Great Britain, we should not only be able to obtain better returns—materially, technically, and financially—from the existing woodlands, but should also in other respects be in a much better position for profitably utilizing some of the existing waste lands for the purpose of entering into direct competition later on with these foreign imports, consisting chiefly of coniferous species of timber ; for these trees can attain as fine development as on the continent of Europe, and often finer, especially in Scotland.   There is indeed no climatic reason why a very considerable portion of the timber now annually imported from Russia, Scandinavia, and Germany, should not in future be supplied of home growth, when once the crops raised have been subjected to *rational treatment from the time of their formation onwards until their maturity.* This latter condition is essential ; for woods that are crowded at thirty, forty, or fifty years of age, may not have been of sufficient or *normal density* at ten or fifteen years of age, but may have become crowded in canopy through excessive and uneconomical ramification and coronal development.

From the *Minutes of Evidence* (pp. 4, 5, and 42) accompanying the Report of the Select Committee on the *Administration of the Department of Woods and Forests and Land Revenues of the Crown*, submitted to the House of Commons on July 26, 1889, it appears that the total area classifiable legally as Crown forests is only 109,139 acres, of which merely 57,304 acres are actually under timber crops, i. e. only 2 % of our three million acres of woodlands are owned by the State, whilst 98 % belong to private landowners.   Comparatively slight though this Crown ownership may appear in amount, yet 57,304 acres, or ninety square miles, represent a very considerable monetary capital invested in soil and timber crops.

Over two hundred years ago John Evelyn, whose classic work *Sylva, or a Discourse of Forest Trees*, was read by him before the Royal Society of London in 1664, bemoaned the great decadence and destruction of the woodlands, or as he termed it ' *the impolitic diminution of our timber*,' that had been handed down to us from primeval times, and ' *which our more prudent ancestors left standing for the ornament and service of their country.*' But he probably little suspected that the very same causes, which had led to what he considered excessive clearance of the remnants of the aboriginal forests, would ultimately lead to their being partially restored by re-planting, that is to say, the desire to get the best possible monetary returns from the soil.

The effect of the gradual depression of agriculture that has been making itself felt during the last fifteen to seventeen years has, of course, first affected the area under the plough. As Major Craigie remarks (*Agricultural Returns* for 1891, p. 10) :—

'Turning to the details of the cultivated area, it is again necessary to note the remarkable changes which have been taking place in the ratio of arable to pasture land in Great Britain. *The two great divisions of arable and pasture now claim, for the first time, an almost exactly equal share of the surface. Twenty years ago the arable land was to the grass as 3 is to 2.* It exceeded by 6,000,000 acres the surface of permanent grass, there being 18,403,000 acres returned as arable to 12,435,000 acres of pasture.'

Again, in the *Returns* for 1892 (pp. 10, 11, 21) he has the same sad tale to tell :—

'*The arable land, as has been the case in every year but two since* 1872, *again shows a reduction.* The surface appearing in this category is 157,000 acres less than in 1891. The permanent pasture in 1892 is also less than that returned in 1891 by 76,000 acres. This is a change in an opposite direction to those recorded for a considerable period, but it is wholly explained by a stricter definition of the term Permanent Grass now enforced in certain mountainous counties, where some of the additions made in 1891 to this category were found on closer inquiry not to have been fully justified, the areas in question 'being again relegated to the class to which they properly belonged, of uncultivated hill grass. . . . .

'Between 1872 and 1882 about 936,000 acres were apparently withdrawn from arable tillage and reappeared in the opposite category of the cultivated area in the form of permanent pasture.

'In the later ten years a similar process has continued. Between 1882 and 1892 the arable area has again diminished, and this time by 1,165,000 acres. . . . The more important alterations between 1891 and 1892, occurring in the entire United Kingdom, may be summarized in the accompanying Table :—

Acreage.	1892.	1891.	1892 compared with 1891.	
			Increase.	Decrease.
	Acres.	*Acres.*	*Acres.*	*Acres.*
Total Cultivated Area . .	47,977,903	48,179,473	. .	201,570
Total of Permanent Pasture	27,533,326	27,567,128	. .	33,802
Total of Arable Land . .	20,444,577	20,612,345	. .	167,768

Owing to the gradual agricultural depression, the poorer classes of arable soil have during the last twenty years gone out of plough cultivation, and have been transformed into pastures; and the immediate consequences of the past year or two must inevitably result in the poorer classes of pasture being ultimately planted up again, thus reverting into something of their original state, as sylviculture will probably hold out better hope of some profitable return from the land than any agricultural or pastoral utilization of the soil[1].

---

[1] During September 1893 the Board of Agriculture drew the attention of landowners to the increased facilities for the planting of woods and trees in

Of course, in such an enlightened country as Great Britain, it might reasonably be expected that everything has already been done by Government to provide for the due instruction of those to whom are confided the responsible task of administering or working these woodlands.   That would certainly be the natural inference when one looks abroad at other countries, and at the same time thinks of the many tens of thousands of pounds spent on, and in connexion with, the fine collections of stuffed birds, fishes, reptiles, &c., palatially housed in the magnificent building forming the Natural History Museum at South Kensington, and on the other vast sums of national money originally and annually disbursed for the housing and maintenance of aesthetic and other technical education.   But one would be vastly mistaken in making any such common-sense assumption.   When Governments are badgered about any subject in an inconvenient manner, they seem to have a stereotyped way of gravely appointing a committee for the purpose of reporting on this, that, or the other subject ; and after such report has been submitted, there is often practically an end of the business in the meantime, until they are worried again on the subject. This solemn farce has been twice recently enacted with regard to Forestry in Great Britain.   During 1885 a committee was appointed for the purpose of considering '*whether by the establishment of a Forest School, or otherwise, our woodlands could be rendered more remunerative,*' and on August 3, 1887, they delivered themselves of a report, from which the following excerpts contain the principal conclusions arrived at :—

Scotland afforded by the *Improvement of Land (Scotland) Act*, 1893, which received the Royal assent on August 24, 1893.   Hitherto owners of land in Scotland have been able, with the sanction of the Board of Agriculture, to charge their estates for the planting of woods and trees, only in cases where the planting is for the purpose of providing shelter.   By the Act in question, this limitation has been removed, and applications may now be made to the Board for sanction to charge estates under the provisions of the *Improvement of Land Act*, 1864, with the cost of planting whether for shelter or otherwise.

'The woodlands belonging to the State are comparatively small, though even, as regards them, the difference between skilled and unskilled management would itself more than repay the cost of a forest school.

.   .   .   .

'Your Committee are satisfied that, so far as Great Britain and Ireland are concerned, the management of our woodlands might be materially improved. Moreover, the present depressed values render economical and skilful management even more important than if the range of prices were higher, though it is probable that, with the waste of forests elsewhere, a brighter future is in store for home forestry, and that *some considerable proportion of the timber now imported, to the value of £16,000,000, might, under more skilful management, be raised at home.*

'Nearly every other civilized State possesses one or more forest schools. In this country, on the contrary, no organized system of forestry instruction is in existence excepting in connexion with the Indian service.

.   .   .   .

'Your Committee recommend the establishment of a forest board. They are also satisfied by the evidence that the establishment of forest schools, or at any rate of a course of instruction and examination in forestry, would be desirable, and they think that the consideration of the best mode of carrying this into effect might be one of the functions entrusted to such a forest board.'

As regards the outlay connected with so flimsy a scheme for the improvement of the natural knowledge of forestry in general, it was also stated that—

'The expense of secretarial staff and examiners need not, in the opinion of the Committee, exceed £500 a year, *and the cost might be considerably reduced by fees for diplomas.*'

I believe I am correct in stating that, up to date, all that has been done by Government for the advancement of forestry in Britain has been to pay £100 a year to Edinburgh University for a lectureship during the last two years, to provide half of the annual stipend of £500 a year to the Professor of Agriculture and Forestry at the Durham College of Science, Newcastle, for the chair founded there in 1890, to pay (temporarily) £150 a year to the free class for foresters, wood-reeves, and gardeners, now being conducted experimentally at the Royal Botanic Gardens, Edinburgh, and to give an equal sum for the elementary course of instruction begun in 1892 at the Glasgow Technical Institute.

The second Committee was appointed in 1889 ' *to inquire into the Administration of the Department of the Woods and Forests and Land Revenues of the Crown,*' and duly delivered itself of a somewhat lengthy report on July 30, 1890, which terminated as follows :—

' *The Committee are of opinion that on the whole the estates are carefully administered,* and that the Commissioners discharge their duties faithfully and efficiently.'

This certainly does not tally with the previous finding of the Forestry Committee, that—

' The woodlands belonging to the State are comparatively small, though *even, as regards them, the difference between skilful and unskilled management would itself more than repay the cost of a forest school.*'

From the evidence adduced before the Forestry Committee it is impossible to deny the truth of the latter statement, and therefore we must look for some explanation of this glaring discrepancy. It is to be found in the fact that the inquiry into the administration of the Crown Woods and Forests and Land Revenues was not at all a satisfactory one, so far as the national Woods and Forests were concerned.

The chief Crown appointments with regard to the administration of the actual Woods and Forests are the following :—

	Salaries.	Allowances.	Total.
One of the two Commissioners of Woods	£1200		£1200
Deputy Surveyor, New Forest, &c. . .	680	£369	£1049
,, ,, Dean Forest, &c. . .	550	304	854
,, ,, { Windsor Parks and Woods } .	500	340	840

Surely the manner in which appointments are made to these chief administrative and executive posts must have enormous influence on the administration itself? Had any body of purely business-men been put to the task of satisfying themselves as to the real nature of the administration of these valuable public estates, they would undoubtedly have asked each of these fortunate public servants what professional education or qualifications he had obtained previous to receiving

his appointment ; for no private landowner would entrust the
management of his estates to any steward or agent without
knowing something about the qualifications and capacity of
the latter.   Yet no inquiries of this description were made by
the Committee ; and appointments to the Department of
Woods and Forests still continue to remain in the patronage
of the First Lord of the Treasury, under Acts 14 and 15
Victoria, chapter 42, section 7, and are in practice more often
filled by the appointment of those having influence with
politicians, than of men having the best qualifications for the
work required of the officers.

Whether or not our three million acres of woodlands can,
by better general knowledge of the natural laws governing
tree-growth, be made to yield $\frac{1}{2}$, or $\frac{3}{4}$, or 1 % beyond what
they now give as a return on the capital represented by
the soil *plus* the growing stock (and the cost of production of
which exceeds $20\frac{1}{2}$ millions of pounds sterling), is surely
a matter of considerable national importance.   It is by no
means advisable, or even justifiable, to pin one's faith to the
Laird o' Dumbiedyke's advice to his son, part of which has
been adopted as the motto of the Royal Scottish Arboricultural
Society :—'*Jock, when ye hae naething else to do, ye may be aye
sticking in a tree; it will be growing, Jock, when ye're sleeping*[1].'
This is in itself fundamentally wrong, for the activity of the
chlorophyll of the foliage, and consequently the most active
assimilation, can take place only under the influence of sunlight ; whilst all our trees have a long period of winter rest—
during which even the evergreen conifers merely transpire
through their foliage *but do not grow.*

---

[1] Waverley Novels, A. & C. Black's edit. 1860, vol. xi. p. 294.   If the
foot-note accompanying the Laird's advice be actual fact, and not also
included in the fiction, then it simply shows how 'enterprises of great pith
and moment' are sometimes lightly embarked on in the crassest and most
unthinking ignorance.   The judicious formation of plantations is no pastime
for an idle holiday ; unless knowledge, circumspection, and thought are
brought to bear on the subject, it bodes an ill omen for the thriving and
ultimate development of the young woods.

But even during the period of active vegetation the trees may perhaps be growing in a way that is not at all economical; they may be dissipating their vital energy in an excessive development of branches and crown instead of conserving and utilizing it for the rapid formation of a longer, straighter, and more full-wooded bole yielding the maximum of technically useful, and at the same time financially profitable, timber. Moreover, whilst that may be a very suitable motto for an *Arboricultural Society*, concerned to a certain extent with the planting of trees for their natural beauty, the full aesthetic effect of which can only be attained by *letting them hang as they grow*, it would be utterly out of place so far as Forestry is concerned; for woodland trees require to be tended, and educated, and carefully ministered to, during all the several stages of their development, in order to produce the largest returns during the various periodical thinnings, combined with the largest and most valuable final yield, materially and financially, when the crop attains its full maturity, and has to be cleared for the reproduction of a younger generation of trees. And can any one in his sound senses think that Forestry, any more than Agriculture, may be conducted profitably and economically merely by following old saws or unthinking rule of thumb? Does the old Horatian maxim, *doctrina sed vim promovet insitam*, not equally hold good with regard to sylviculture, as well as to other arts and sciences? Is Forestry so essentially different in its natural laws and fundamental principles from Agriculture, that whilst University chairs have been founded for the latter at Edinburgh, Belfast, Newcastle, and in a manner also at Oxford[1], and properly equipped Schools of Agriculture are located at Edinburgh, Carlisle, Cirencester, and Downton, absolutely next to nothing has yet been done for the dissemination of sound and properly qualified instruction in sylviculture?

[1] The Sibthorpian Professorship of Rural Economy at Oxford is at present suspended, but has hitherto been held only by an Agriculturist.

It may be urged against this, that at some Agricultural Schools there are classes of *Estate Management and Forestry.* This is quite correct, but it does not follow as a matter of course that the teachers there have a really sound knowledge of the scientific principles and practice of Forestry. Hence a question may very pertinently be asked similar to the *Quis custodiet ipsos custodes?* For where shall such men at present get the scientific and practical training in Forestry, without which they are in a certain degree like the blind leading the blind? In order that this picture may not seem harsh, ungenerous, and overdrawn, permit me to quote from the evidence given before the parliamentary *Committee on Forestry* on June 15, 1887, by the lecturer on Estate Management and Forestry at Cirencester (pp. 27–29 of Report dated August 3, 1887).

468. You are Professor of Estate Management and Forestry at Cirencester College?—I am; I am also a Fellow of the Royal Highland Agricultural Society of Scotland, by examination.

470. Have you had practical instruction in forestry yourself?—No, only on estates.

474. Would you give the Committee *some idea of the course of instruction in practical forestry?—There is a course of lectures forming a part of the syllabus of estate management, consisting of about six or seven lectures,* as the case may be, and those are illustrated by field-classes in the woods, timber measuring, and valuing standing timber, different processes of planting, grafting, layering, and so on. This is done in the woods.

485. And are they (the students) shown specimens of good forestry and bad forestry, and how to distinguish good from bad?—We have not that opportunity; *we have no bad forestry.*

487. You show them what mistakes to avoid in wood management I presume?—Yes, but at present it is only on a very small scale in my syllabus; we have not an opportunity of going fully into the matter.

509. I gather, from what you have stated, that you do not consider the course at Cirencester is sufficient for the education that land agents should receive in forestry?—No.

515. Is it your belief that instruction in forestry would be a national gain?—It is so.

521. *Are you acquainted with the system of forest instruction pursued in France and Germany?—No, I am not.*

527. You were taught yourself at Cirencester College, were you not?—Yes. (Compare 509.)

528. And after then where did you go?—I went into King's College in London for twelve months, and then up to Edinburgh, and took the Highland Agricultural Society's Diploma; that was about the year 1868.

529. *So that you have had yourself as good an education in forestry as is possible in Great Britain?—There was no teaching in forestry in my time.*

530. *But you have had a good insight into forestry?—I have had an insight into forestry.*

541. Do you think it would have been an advantage to yourself if you had had an opportunity of being trained in some regular school, or do you think you have been able to acquire a practical knowledge?—I have been fortunately situated with regard to that. I have had a very long and happy experience with regard to forestry; but others who might follow in my steps might not have the same opportunities of instruction, and others ought to have the opportunity of hearing lectures upon the matter.

Nothing that is said above is meant, *malo grato*, to depreciate the instruction given in Forestry at these Agricultural Colleges. But the evidence tendered by one of the instructors himself is plain and to the point, as it shows most conclusively that the instructors in Forestry have not themselves had good opportunities of qualifying for their special task, so far as it relates to Forestry conducted on sound scientific and economic principles, i. e. to *Sylviculture* as contrasted with *Arboriculture*.

Perhaps the only place, with the exception of Cooper's Hill, at which really properly qualified instruction in Forestry is being given at the present moment is the Durham College of Science at Newcastle. But as the lectures on Forestry there only form a minor portion of the course of professorial tuition dealing principally with Agriculture, they can hardly be expected to be adequate to the proper qualification of youths for the management of our existing woodlands in full accordance with the principles and the practice of modern scientific Sylviculture as fully expounded and demonstrated in Germany, the home of Forestry in its highest technical and practical perfection.

The course of lectures which was organized in 1889 at Edinburgh University is, as at present constituted, merely a makeshift; and even as a temporary compromise it is only satisfactory on the broad principle that ' half a loaf is better than no bread.'   At the present moment reference is made only to a course of technical instruction fitting young men of liberal education for the management of estates containing a considerable area of woodlands.   It does not take into consideration the wants of that other very important class of less highly educated men occupying the position of foresters, wood-reeves, bailiffs, overseers, &c., for whose improved education an experimental system of instruction has recently been inaugurated by Professor Balfour at the Botanic Garden in Edinburgh.   For the ultimate success of both of these Schools every lover of Forestry and of woodland growth must hope.

The only educational institution properly equipped with duly qualified teachers for granting a sound Sylvicultural education at the present moment is Cooper's Hill College, near Staines in Surrey, originally built and fitted up as an Indian Engineering College at a cost of £134,000.   Without considering the causes which led to the necessity for terminating the original course of study of the probationers for the Indian Forest Service in Germany and France, and without discussing whether the cessation of the continental training has proved itself beneficial or not, it may be said that Cooper's Hill was not primarily selected for any local or other advantages it offered *per se*.   Its selection was mainly due to other reasons; because it was an expensive establishment already in the hands of the India Office, and needed financial support after the reduction in the number of young officers annually required in the Public Works or Civil Engineering Department (in consequence of the fall in the rupee), for the education of whom alone the maintenance of the College had become unduly expensive.   Previous to the education of the probationers for the Indian Forest Service being begun there in 1885, an annual deficit had for some years to be met; but by arranging for the education of these, and charging them the usual College fees of £183 a year, this deficit was transformed into a surplus,—thereby proving that the sum charged is in excess of what it actually costs to feed and educate the lads.   Had the large sums of money which are being granted annually as salaries to the teaching body there—which consists of three Professors, three Lecturers, and one Instructor on the Forest branch of the College Staff, with the additional assistance of one Director of Practical Study on the Continent, who superintends the course of study and the tours of inspection made throughout different parts of Germany in order to see practical work on a large scale—been devoted to the foundation of two chairs, one of Sylviculture and Management of Forests, the other of Protection of Forests and Utilization of Forest Produce, at each of the three Universities, (1) Oxford or Cambridge, (2) Dublin or Belfast, and (3) Edinburgh, or perhaps better still St. Andrews or Aberdeen, from their proximity to extensive forests, Great Britain might now have had a course of sound scientific instruction at central points in England, Ireland, and Scotland, fairly abreast of continental Forestry in so far as our own particular requirements are concerned.

The present instruction in Forestry at Cooper's Hill can never be regarded as capable of being converted, without previous translation to some University or Agricultural College, into a national place of education in Sylviculture. Because, in the first place, the lectures on Forestry are not capable of being listened to *ex viâ* along with any simultaneous course of instruction in what may be more likely to prove better *bread-winning studies* (although at a University this might very easily be arranged for); whilst, in the second place, the charge of £183 a year, and a three years' course, make such a training absolutely prohibitive for the majority of those who otherwise might profit by a shorter and less expensive course of instruction; for, after reckoning all extras, clothing, travelling, and various incidental expenses, it must cost a parent or guardian at least £700 to £800 to provide his son or ward with a three years' course at Cooper's Hill.

This is, however, not a matter that really concerns British Forestry; for Cooper's Hill is maintained by the India Office, and if the Government of India choose to maintain an expensive institute for young men who might easily be educated at one or other of the great Universities, it is no direct concern of the British public, and is, at any rate in the meantime, beyond the legitimate sphere of their control.

There are three distinct classes for which it has, face to face with the existing decline in the agricultural and pastoral value of many kinds of land, now become really of national importance that some scientific instruction in Forestry should be provided; these classes are:—

I. The owners of woodlands, or of waste lands, or of lands of poor quality now likely to be soon thrown out of arable or pastoral occupation.

II. The higher educated class which supplies the land-agents, estate factors, stewards, and the like.

III. The class of smaller stewards, bailiffs, foresters, wood-reeves, overseers, &c.

There is no need to prove that each and all of these classes really require instruction in Forestry. It has been often admitted, and is apparent throughout both of the Forest Reports previously quoted. Even should a man be born to inherit broad acres of stately woods, it does not follow that he is naturally endowed with an intuitive knowledge of the laws of nature regulating the normal development and the best technical methods of tending woodlands; whilst estate agents and managers may be satisfied with results, that would perhaps not seem to them the best attainable if they were better acquainted with the principles and practice of scientific Forestry. The necessity for instructing landowners is nowhere more apparent than in Sir Herbert Maxwell's paper on *Woodlands* in the *Nineteenth Century* for July 1891, in which in less than a dozen lines he points out one great blemish in our method of treatment of woodlands, and then at the same time, in tendering advice on the subject, commits himself to the expression of an opinion which distinctly shows that his ideas on Forestry have no true scientific foundation and are opposed to the very best and most thorough natural knowledge on the subject[1]. He boldly advocates the formation of *pure forest* in preference to mixed woods. I am, however, perfectly certain that, if once the formation and retention of mixed woods be departed from on any extensive scale, the woodland owners throughout Britain will begin to form an intimate and unpleasant acquaintanceship with many noxious insects, some of which, rejoicing in the euphonious names of *Gastropacha*, *Liparis*, *Tomicus*, *Pissodes*, *Hylesinus*, &c., will give ample reason for genuine regret at having transformed mixed woods into pure forests. And there are many other objections that can be named; but this is not the proper time and opportunity for doing so.

The necessity for men of the second class receiving a better instruction is also everywhere apparent throughout the Report on Forestry. It may be summed up in the quotation of two answers given to the Committee by Mr. Britton, the manager and valuer of a large timber-firm in Wolverhampton :—

783. Do you find that many land-agents possess a capable knowledge of forestry ?—Very few in all my experience ; *I think I could pretty well count them upon my fingers' ends.*

829. What is the general result that you have come to ?—The general result I have come to is, that *very few land-agents know anything of forestry*, or very little.

And the same holds good with regard to Scotland, although there are a few really good practical foresters holding charge of important wooded estates. Thus the late Mr. Macgregor

---

[1] See Chapter VI, *On Mixed Woods and their Advantages over Pure Forests.*

of Dunkeld, the Head Forester of the Athole woods, extending to over 20,000 acres, gave the following evidence :—

877. Do you consider that the land-agents, or the factors, as they are called in Scotland, are fairly well informed as to the management of woods and timber ?—Very few of them, I think.

878. What, in your opinion, are the subjects on which factors and woodmen are deficient ?—*They are deficient in the knowledge of what trees ought to be planted on suitable soils, and when thinning out ought to commence; and, in fact, the general management of the woods altogether.*

And if the majority of the men concerned with the administration of woodlands are thus ill-provided with knowledge respecting the *formation*, the *tending*, and the *management of timber crops*, it is certainly open to reasonable doubt if they can possibly be in a position to have the soundest ideas with regard to the *utilization* of the latter when they have reached their financial or their economic maturity.

It is, of course, by no means improbable that a shrewd, observant, practical forester may often arrive at proper methods of treating certain crops without at the same time having any true scientific knowledge of the natural laws influencing either tree-growth in general, or the various individual species of trees · in particular.   Mr. Macgregor again spoke with no uncertain voice on this point :—

1047. You have had no instruction (on forestry) ?—No.

1048. *Do you think you would have done better if you had had a course of scientific instruction ?—I have not the slightest doubt of it; I have felt the want of it all along.   I had to read up, and there are very few books to read.*

1049. That is your own experience, and *you are prepared to recommend that men beginning life as foresters should have some definite instructions ?— Certainly.*

The preparation of text-books on the various branches of Forestry was to have been one of the functions of the Board of Forestry recommended by the Committee in 1887.   But practically nothing has been done in this direction.

It would not be at all right if at the same time I did not point out that evidence was given before the Committee on Forestry tending to show that planting up in England could

not promise remunerative returns, owing to the fact of the
timber marts being glutted with foreign timber, to the general
depression in trade which was then making itself felt, and to
the use of substitutes in many ways in which timber was
formerly employed.   But it is important to note that among
Messrs. Thompson, Macgregor, and McQuorquodale, the able
practical head foresters then in charge of the Strathspey,
Athole, and Scone forests, there was a consensus of opinion
that properly managed woodland crops are undoubtedly profitable.   On this point the late Mr. Macgregor's evidence may
be briefly quoted :—

1122. Then you think the forest area in Scotland might be largely
increased?—Very largely ; it can be very much extended.
1123. With profit?—With profit.

This practical advice becomes of all the greater importance
when collated with my previous statement relative to the
gradual *appreciation* which can be proved in the case of timber
up till now.   Simply as a commercial speculation, timber can
and should be grown to a much larger extent than at present.
If any one studies the relative and steady appreciation of timber
in the past, it requires no subtle arguments to make out a fair
case for the judicious sylvicultural utilization of the soil in
many parts of our island.

I do not venture to make this indictment against the
Governments of late as well as of previous years, of being
utterly neglectful of really important national interests relative
to the already existing three million acres of woodlands, without being in a position to offer some practical suggestion for
the removal of this blot on our system of technical education.

But though I cannot agree with the Committee on Forestry
in attempting to remedy the defect for so small a sum as £500
a year (reducible by fees for diplomas), yet it need cost a mere
nothing in comparison with the housing and maintenance of
our national Natural History Museums at Kensington and
elsewhere.   Sir Herbert Maxwell, in his paper on *Woodlands*

already referred to, stated that if the landowners of Britain really wanted instruction to be provided at any educational centre or centres, they could easily arrange for it among themselves without the necessity of appealing to aid from the State. From one cause or another, though probably not that they are already perfectly satisfied with the existing state of affairs, the landowners have failed to subscribe sufficient for even half the amount necessary for the endowment of one rather poorly-paid Chair of Forestry at Edinburgh. Despite the efforts which have continuously been made in this direction by the Highland and Agricultural and the Arboricultural Societies of Scotland, the Secretary of the former had to report to the Council at its meeting on November 2, 1892 :—

' that he had sent out over 5,500 circulars requesting subscriptions for the endowment of the Chair of Forestry in the University of Edinburgh. He regretted that the result had been disappointing, but that the sum now subscribed from all sources amounted to over £2,300.'

This apathy on the part of landowners does not, however, justify any *laissez faire* policy on the part of Government; on the contrary, it should make the latter all the more anxious to remedy the educational defect for the benefit of the rural working-classes, and of our timber-consuming industries, both of which would be very directly and materially benefited.

To commence with the lowest scale of requirements, those affecting the forester or wood-reeve class, the experimental tuition now being given under Professor Balfour's auspices at the Botanic Garden and Arboretum in Edinburgh is a step in the right direction. But whilst it affords the very best opportunities of providing lectures on elementary scientific subjects affecting Forestry, it fails in one essential to practical success ; for the work the men are engaged on, whilst going through the course, is, and must continue to be, either *Horticulture* or at best *Arboriculture*, but is certainly not *Sylviculture or Forestry*. If we were to follow the excellent Bavarian example—and I do not think we could do better—we should establish

Sylvicultural Schools for young lads at Dunkeld in Athole, or Grantown in Strathspey for Scotland, and at Coleford in the Forest of Dean or Lyndhurst in the New Forest for England, where it would be easy to provide a fair practical education for a limited number. Of course in this event there would not be the scientific teachers within easy reach to give short courses of elementary instruction; but a little extra payment for travelling expenses might no doubt circumvent the difficulty, and overcome that drawback.

For the first and second classes, those of landowners, land-agents and factors, nothing short of the establishment of Chairs of Forestry at Universities seems in any way to meet the necessities of the case; for, as I have already asserted, it is only at such educational centres, and not at any special colleges, that an opportunity can be given to students of hearing lectures on this subject concurrently with other and more usual studies fitting them for the ordinary vocations in life. And perhaps, taking all things into consideration, the most suitable Universities for the dissemination of scientific instruction would be Oxford for England, Edinburgh, and either St. Andrews or Aberdeen for Scotland, and Dublin or Belfast for Ireland. Even if Government were willing to establish Chairs of Forestry at any or all of these places, it by no means follows that the Universities applied to would accept the proposals, and permit the incorporation of such Chairs into the existing teaching body, unless they were granted a perfectly free hand with respect to the selection and appointment of the Professors.

If any such Chairs be formed from national funds, then the salaries attached to them should be sufficient to obtain the services of really highly-trained men, and should not be less than £700 a year at Oxford, Edinburgh, and Dublin, or £600 at St. Andrews, Aberdeen, or Belfast; and the lectures should be free to all who wish to hear them—whether matriculated students of the year or not. The reason of this will be apparent when it is stated that the attendance given to the lectures at Edinburgh was 7 in 1889–90, 6 in 1890–91, 40 in 1891–92, 10 in the winter of 1892–93, and 5 in the summer of 1893, during the first, second, and fourth of which sessions the fee for the course was £3 3*s.*, whilst it was free (to matriculated students only) for the merely incomplete course given during the winter session of 1891–92.

If formed, none of the appointments should be filled except by men thoroughly trained in Forest science, and fully qualified to give lectures over the whole range of Forestry ; that is to say, in—

1. *Sylviculture*, treating of the formation, tending, and reproduction of woodlands.

2. *Protection of Woodlands*, showing how to safeguard them from inimical inorganic agencies (winds, frost, &c.) or organic enemies (fungoid diseases, insects, game, &c.).

3. *Management of Woodland Estates, and Valuation of Timber Crops*, for the framing of Working Plans, and in order to determine the most profitable methods and times of utilizing the timber crops.

4. *Utilization of Woodland Produce*, showing how it can most economically be prepared for, and brought to, market.

The matter is either one of true national importance, or it is not. If not, then the owners of woodlands may well be left to go on as indifferently and unintelligently as they have hitherto been doing ; for they can help themselves out of the difficulty whenever they choose to combine and open their purse-strings to no very great extent. But if it really be of national importance, then now, when large areas of land are likely to be thrown out of arable and pastural occupation, is the proper time for the Government to initiate a sound system of instruction in Forestry. Even if my suggestions were carried out to their very fullest extent ; if Chairs of Forestry could be founded at Oxford, Edinburgh, Dublin (each at £700 a year), and at St. Andrews, Aberdeen and Belfast (each at £600 a year), and if small schools of Practical Sylviculture for youths were arranged for at Dunkeld, Grantown, Coleford, and Lynd-hurst (at about £150 a year each), the total annual expenditure involved would only amount to £4,500. This is surely a very slight insurance to pay for the probable better management of woodlands already amounting in actual minimum cost to over twenty-and-a-half millions of pounds sterling, and most likely to be very considerably added to in the immediate future. It amounts, in fact, to less than three farthings per acre per annum for the land already actually under timber.

CHAPTER II

THE BRITISH SYLVA, AND THE GROWTH OF WOODLAND CROPS IN GENERAL

AT the time of the Roman invasion of Britain a very considerable portion of our island was still covered with forest, as the gradually advancing requirements of increasing population had not yet necessitated any very extensive clearance of the primeval woodlands in order to permit of the expansion of tillage and pasturage.

The species of forest trees of which the woods consisted comprised Oak, Beech, Hornbeam, Scots Pine, Birch, Ash, Willow, Alder, Scots Elm, Yew, Aspen, Mountain Ash, and Hawthorn. The uplands of central and southern England, and more especially those tracts having limy or chalky soils, bore dense woods of Beech (which, however, was not indigenous to Scotland, where it was not introduced until the beginning of the eighteenth century), whilst a stately growth of Oak covered the richer alluvial stretches having deeper soil. The sandy hills of southern England and all the mountainous tracts extending from Yorkshire to the northern portion of the island were probably mainly covered with Scots Pine (the only species of the *Abietineae* indigenous to Britain), Birch, and Mountain Ash, whilst Oak, Ash, Scots Elm, Aspen, Willow, Alder, and Yew were for by far the most part restricted to the dells, valleys, and lower-lying localities, where the soil was generally deeper and more productive, and the climate was less rigorous.

During the Roman occupation of Britain (A. D. 44–410) many species of trees were introduced from the continent of Europe, including Sweet Chestnut, English Elm, Poplar, and Lime, together with a good many other ornamental and fruiting trees which failed more or less to establish themselves, and never actually attained anything like true forest growth. Even these four species named have only naturalized themselves to a certain degree; for they cannot in our climate, and more particularly in that of the northern portion of the island, with its shorter and considerably cooler summer as compared with the south of England, be relied on to produce seed of normal germinative quality. To compensate for this, however, they are richly endowed with the power of throwing out stoles or suckers from the roots, and these are capable of being transplanted like seedlings.

Of the various other species that are now entitled to rank among our forest trees, Sycamore or Scots Plane, white and crack Willows, and white and grey Poplars were introduced before the end of the fifteenth century; Spruce, and Cluster Pine during the sixteenth century; Silver Fir, Maple, Horse Chestnut, and Larch during the seventeenth century (although the Larch was not introduced into Scotland till 1727); Weymouth, Maritime, Cembran, and Pitch Pines in the eighteenth century ; and Austrian, Yellow, and Jeffrey Pines, Nordmann's and Douglas Firs, Menzies Spruce, &c., during the present century.

The Forest Trees of any country may be considered, from a sylvicultural point of view, as belonging to one or other of two classes ; for they must be either **Principal** or **Ruling Species**, forming, or capable of forming, pure forests without any assistance from other kinds, or else **Subordinate** or **Dependent Species**, which are practically found thriving best in admixture with one or more of the ruling species.

The terms **principal** and **subordinate** are of course here used

in a purely sylvicultural, and not in any financial, sense ; for, as
a matter of fact, the most valuable species of timber are usually
to be found among the latter class.   To the former belong,
among the conifers, Scots Pine, Spruce, Silver Fir, and occasionally Larch, and among the broad-leaved trees, Beech, Oak,
Birch, and Alder principally; whilst the latter include all the
other indigenous and exotic trees now of woodland growth in
Britain, of which the principal are the Douglas Fir, the Menzies
Spruce, the Black Pines (Austrian and Corsican), and the Weymouth Pines among conifers, and the Ash, Maple, Sycamore,
Elm, Hornbeam, Aspen, and Willow among broad-leaved species.

For the sake of brevity the following details will be confined to the above-named trees only.

Whilst in growth trees extract supplies of nutriment partly
from the atmosphere (in the shape of carbonic acid, the
carbon being assimilated, and the oxygen being set free) and
partly from the soil occupied by their root-systems (mineral
nutriment).   If the trees were to be allowed to die off and
then rot, in place of being utilized as timber whilst still in healthy
growth, the supplies of mineral nutrients extracted from the
soil would be restored to it, by the formation of humus or
mould, on the ligneous tissue becoming decomposed under the
action of sunlight, moisture, and a moderate degree of warmth.
But, with the removal of the timber crops, the natural balance
is disturbed ; and, unless protection be afforded to the agencies
active in causing the chemical changes that result in the decomposition of its mineral constituents, the soil runs a risk
of becoming gradually deteriorated and finally exhausted.   In
agriculture, a similar danger is obviated by manuring the land,
the manure not only helping to replace the nutrients previously
withdrawn, but also stimulating to increased nitrification and
decomposition of the soil.   In sylviculture, however, such
measures would be neither practicable nor remunerative ; and
the only recompense that can practically be given consists in
the fall of dead foliage, twigs, &c.   Fortunately, however, in

these débris is stored up a large portion of the mineral substances extracted from the soil. But the work of decomposing and dissolving the mineral nutrients is indirectly aided by the leafy canopy of the woodlands. This both prevents insolation of the soil, and also safeguards the retention of a certain amount of soil-moisture requisite for rendering the mineral nutrients soluble; because it is only in the form of soluble salts that food can be imbibed by the rootlets. By ramifying throughout the soil, the root-systems help to cleave and fissure the subsoil previous to other changes taking place; whilst shrubs, grasses, and mosses, thriving under the shade of the trees, also die off, decompose, and are transformed into forest mould, which, from its strongly hygroscopic nature, absorbs and retains the atmospheric moisture and precipitations, and assists in regulating the flow of water within the soil.

When the **leaf-canopy** overhead is much interrupted or broken, the soil is apt to be overrun with a rank growth of noxious grasses and weeds, which consume the nutrients unprofitably; whilst at the same time it is exposed to the inimical influences of sun and wind, whose action is apt to reduce the quantity of moisture in the soil below that which is most favourable for tree-growth.

For the retention of the soil-moisture, and the protection of the soil generally, those species of trees are the most favourable crops, which have a dense crown of foliage and maintain themselves in close canopy until their maturity—qualities with which the different kinds of trees are very unequally endowed. Among broad-leaved genera the Beech ranks highest in this respect, its canopy being close and its foliage dense, whilst the dead leaves are rich in potash and form mould of excellent quality and strongly hygroscopic power. The leaves of Oak do not form good humus, as a considerable amount of tannic acid is evolved during decomposition. Hornbeam foliage yields good mould, but the crop of leaves is lighter in every way than in the case of the Beech; Lime shades the soil well and is thickly foliaged, though, as the leaves are thin in texture, the amount of humus formed is only slight ; the Chestnut has a thick crown of foliage yielding good mould. None of these last three genera occur, however, to any great extent forming pure woods in Britain. Oak, Elm, Ash, Maple, Sycamore, and the other valuable broad-leaved species are, owing to the comparative lightness of their crowns of foliage, not naturally well endowed with capacity for protecting the productive power of the soil, when they form crops by themselves, without the assistance of other trees with denser crowns,—although they may be found doing well on soils of so moist a character that insolation and evaporation are rather beneficial than prejudicial in diminishing the amount of soil-moisture. Among conifers the species best endowed with soil-improving and soil-protecting qualities are in particular Spruce, Menzies, Douglas, and Silver Firs, and in a less degree the Weymouth and the Austrian and Corsican Pines, the fine growth of mosses beneath which also acts like humus in sponge-like absorption and retention of moisture. Scots Pine and Larch are, owing to the sparseness of their foliage, least endowed in this respect, although during the earlier stages of their development a good layer of moss is usually to be found covering the soil, and protecting it from the inimical and exhausting effects of sun and wind. When once the canopy begins, however, to get broken, this beneficial covering of moss disappears ; and then the soil, with its new growth of rank grasses, whortleberry, or heather, rapidly begins to deteriorate, unless underplanting is taken in hand. Where marshy land is planted up with Alders, Birch, Willows, and Poplars, the action of sun and wind is often beneficial, just because it tends to reduce the quantity of moisture, which is apt to be excessive on such low-lying situations.

One of the leading principles in the Management of Forests is that in woods of normal composition, having crops of all ages and classes properly represented from zero up to maturity on areas of equal productive capacity, the annual fall of mature timber shall, under ordinary circumstances, be equal to the annual production of the total area during one year; and this is represented, as the total growth throughout x years, by the mature timber standing on the fall about to be harvested. To cut down more than this, is to consume a portion of the capital represented by the growing stock; whilst to utilize less, is simply not to treat the woodlands in an economic manner. In order, however, that such normal average increment may take place annually, it is necessary that the woods must be of **normal density**, that is to say, neither too crowded on the one hand, nor too sparsely stocked on the other, although of course the suitable and proper degree of density that may be termed *normal* varies with each species of crop. Crowding is disadvantageous. If the individual trees forming the crop have not a sufficiency of growing-space, then they cannot possibly obtain the supplies of light, air, and warmth requisite for their most favourable development, and therefore soon fall into a sickly condition of growth, which predisposes them to dangers from wind, snow, insects, and fungoid diseases. The constant struggle, too, which is being waged for supremacy among the individual trees in woods becomes thereby unduly prolonged; and in this is dissipated the energy which, with judicious tending and thinning, can be advantageously utilized for the better and more vigorous formation and development of the individual poles or trees of more robust growth, from among which the trees finally forming the mature crop must ultimately be selected. But if the number of individuals per acre falls below the normal average density for any particular species of tree and any given class of soil and situation, or, in other words, if the growing-space allotted to each individual be in excess of what is requisite to maintain the pole or tree in vigorous growth, then its vital energy becomes dissipated in branch-formation, instead of being conserved so as to be concentrated and utilized in the formation of a long, straight, full-wooded bole or stem, free from knots, and consequently of higher technical utility and commercial value.

And, at the same time, from the leaf-canopy formed by the crowns being much more apt to be interrupted here and there, or sometimes quite broken in its continuity, the surface of the soil must be exposed to the action of sun and wind and to the consequent danger of rank growth of weeds, subsequent deterioration, and perhaps final exhaustion.

This **normal density**, which should be maintained unbroken or uninterrupted in order that the productive capacity of the soil may be fully utilized, and yet at the same time protected against deterioration, consists of a full or close **leaf-canopy**. The different species of forest trees are very unequally endowed with capacity to retain close canopy when once they have entered on the later stages of development, for they make different demands with regard to individual growing-space, and then the density of the foliage forming the crown also varies with each individual species. To estimate these differences it is only necessary to compare the crown of a Beech or a Silver Fir with that of a Birch or a Larch, although in these cases extremes are purposely chosen in the two classes of broad-leaved trees and conifers. This physical fact, that all trees of forest growth are not alike endowed with the capacity of shading the soil and protecting it against the exhausting influences of sun, wind, and rank growth, is a matter of very great moment from the sylvicultural point of view. Although all our forest trees may be grown in pure woods, if worked with a low rotation, owing to timber crops being naturally dense during the thicket period (that is, till the crop begins to clear itself naturally of dead branches), and during the pole-forest stage of growth (which means till the leading stems have attained a girth of about two feet measured at breast height), yet after that, when once they have entered into the tree-forest or high-forest stage of development, their natural individual differences with regard to demands for light, i.e. growing-space, and with respect to density of foliage in the interior of the leafy crown, begin to make themselves unmistakably apparent. Thus whilst thickly foliaged kinds of trees, like Beech, Silver Fir, and Spruce, can be grown in pure forests until they attain their full technical and mercantile maturity, without the productive capacity of the soil being endangered, it would be contrary to one of the leading principles of sylviculture—viz. the conservation of the productive capacity of the soil—to cultivate pure high forests of **light-demanding genera** like Oak, Ash, Maple, Pine, or Larch, except under special circumstances where the productive capacity of the soil is not endangered by insufficient cover, as, for example, on some classes of marshy land where the evaporation caused by insolation and by the free play of winds is directly beneficial, or on low-lying tracts with fresh soil, whose depth and porosity might perhaps be injuriously affected by any accumulation of humus. It is solely with regard to the soil-protecting capabilities of the different forest trees that the classification above given of **Ruling** and **Dependent Species** is based; for it may be briefly stated that the productive capacity of woodland soil is only safeguarded to the necessary extent when the timber crop consists either of thickly-foliaged species growing in close canopy, and providing the soil with a good layer of leaf-mould, or else of conifers having evergreen foliage, under which a covering of mosses performs the functions of the humus or mould elsewhere. The species of trees naturally fitted to be grown in pure forests under the first of these conditions are pre-eminently Beech, Spruce, Menzies, Douglas and Silver Firs, and in a less degree Hornbeam, Lime, and Chestnut; whilst those falling under the second condition comprise the several varieties of the Pine genus, so long as they are not worked with too long a period of rotation, i. e. so long as their fall is not delayed too long after the time when they sink so far below the normal density of canopy that the growth of mosses gives place to weeds and berries, and the soil begins to deteriorate through insufficient protection against insolation and exhausting winds. Unfortunately the broad-leaved trees best qualified by nature for the formation of pure forests are not in Britain of sufficient technical and commercial value to lead to their cultivation on any large scale; but among them the Beech and, on moister varieties of soil, the Hornbeam are of great sylvicultural importance as the ruling species or matrix, either in the shape of underwood or of tree-forest, along with which the more valuable classes of timber trees, Oak, Ash, Maple, Sycamore, Birch, &c., may be most profitably grown on localities where the soil would be liable to deterioration if the woods consisted of these thinly-foliaged and light-demanding species only.

For the formation of woods of normal density or full leaf-canopy the number of individuals per acre varies, as has already been stated, according to the species of tree; but wherever reliable data have been collected, it appears that, on soils of the best quality for the requirements of each genus of tree, at the age of 120 years there are 37 % more Beech, 24½ % more Silver Fir, and 60% more Spruce than individual stems of Scots Pine per acre. It also appears that at all ages there is on soils of the best quality invariably a very much smaller number of individual trees of any particular species than on the poorer classes of soil. The difference may often be enormous, for at eighty years of age Scots Pine woods on the best class of Pine soil consist of only one-third of the number of trees that are generally found on the poorest class of soil. So far as these different results are concerned, with one and the same species of tree, they are explainable through the fact that the general development and the natural process of selection of the fittest—the struggle for light and air, for domination, and for individual existence—is more energetic and decisive on good soils than on those of inferior quality. The differences exhibited by the various species of trees, however, is dependent on their different sylvicultural characteristics with regard to the natural development of their crowns, and their special requirements as to growing-space. Throughout the Spessart forest it has been found that the following average density of crop is necessary for the formation of close canopy :—

Age of Forest, years.	Beech.	Oak.
	Soil of best Quality.	Soil of best Quality.
60	587	1,234
120	210	326

Although at first glance this almost looks as if it might be argued that the Oak is more capable of bearing shade than the Beech, which would be contrary to experience, it actually proves nothing more than that the Oak takes considerably longer than the Beech to develop in that locality, and that the Beech finds lateral expansion of its crown a *sine qua non* of existence at an earlier period of age than the Oak.

Whilst the number of individual poles or trees forming the crop is larger on poor than on good soils, yet the full normal density of leaf-canopy is maintained longer on the better classes of soil. But the actual extent to which the ground is overshadowed varies also with the species of tree; for experiments and measurements have shown that in a Spruce forest, of normal density for Spruce, the soil is overshadowed to more than twice the extent that obtains in a Scots Pine forest, of normal density for Pine.

If what has been stated above be correct—viz. that the productive capacity of the soil can be best protected when close canopy is maintained throughout woodland crops—then it must naturally follow that the greatest production of timber per acre must take place in woods having a leaf-canopy of normal density. And practical experience shows that such is the case. In general, other things being equal as to the productive capacity of the soil, the increment per individual stem is greater in trees revelling in the full enjoyment of light, air, and warmth, than in those whose crowns are limited to a more or less restricted growing-space; because they have larger root-systems, can draw greater supplies of nutriment from the soil, and enjoy better opportunities for carrying out assimilation and for utilizing their nourishment in the formation of ligneous tissue. On soils similar as to quality the total production per acre in clóse-canopied forest is not only greater than in the more open forest, but the technical quality of the timber produced is also much higher owing to the better dimensions attainable by the bole or stem, and to freedom from knots formed where the branches emerge from the trunk of the tree. The larger the growing-space allowed to each individual tree, the greater is the tendency to branching growth taking place at the expense of the bole; hence, for the production of long straight stems, it is necessary to maintain as close canopy as is consistent with the general health and vigour of the crop until it has passed through the pole-forest stages, during which its growth in height is most active, although this again varies according to species of tree and other factors. When crops attain their chief growth in height after being maintained in canopy of normal density, experience has shown that the stems rapidly increase in girth when the canopy is interrupted, artificially, for the purpose of permitting the individual trees to obtain larger supplies of light, air, and warmth. This sylvicultural operation, although purposely diminishing the number of individual trees per acre, often leads to a greater increment and a more valuable yield than if the crop had been allowed to remain growing in normal canopy,—but this is a subject that will receive full attention subsequently (see Chapter X). As, in woods growing in close canopy, the crown is circumscribed, and can for the most part obtain only from above the light and warmth necessary for the work of assimilation, the poles compete with each other for a larger measure of exposure to these invigorating influences, so that the growth in height becomes stimulated. And as, owing to the crowns being limited to the upper portion of the stem, a larger proportion of the assimilated food is utilized in the upper part of the tree than in the lower, the result is that the boles are more full-wooded, i. e. they approach more to the cylindrical in form, than is the case with poles growing in crops below the normal density in leaf-canopy.   Until the chief growth in height has been completed, therefore, the maintenance of close canopy affects not only the actual length of the bole or clean stem and its shape, but also the proportion of timber obtainable, as compared with branchwood; and these are matters of great technical and financial importance.

When seedling growth or young plantations, after having passed through the thicket stage of growth, develop into pole-forest, and finally into tree-forest or high-timber, the crop consists of various classes of growth named, according to the position they may occupy in the struggle for increased growing space: (1) **predominating**, (2) **dominant**, (3) **dominated**, and (4) **suppressed**, although of course there are no hard and fast lines existent between these different classes.   Until far on in the life-history of the crop, the predominating class comprises within it all the individuals that are likely to form any portion of the mature fall of timber; hence, in all clearings and thinnings undertaken with a view to the tending of woodlands— a subject that will be considered later on (see Chap. IX)— the removal of the suppressed and, to a greater or less extent (according to the species), the dominated classes, is advisable. Such cultural measures are not only for the purpose of assisting nature in the work of eliminating the individuals no longer required in the maintenance of a leaf-canopy of normal density, and of helping the more vigorous stems towards their more rapid development, but also for removing all poles and young trees of enfeebled energy and perhaps sickly growth, which might offer dangerous attractions to noxious insects. Compared with the total number forming the crop, the percentage of individual stems necessary for the retention of a leaf-canopy of normal density varies essentially according to (1) the species of tree, (2) the nature of the soil and the situation, and (3) the previous treatment that has been accorded to the crop.

In consequence of the competition for possession of the soil, which takes place when different species of trees have been intermixed either naturally or artificially, the law of nature regarding the survival of the fittest has brought it to pass that, wherever they have long had free scope to make their special characteristics and influence felt, certain species of trees usually predominate numerically over large tracts of country. As has already been mentioned, the primeval woods of Britain consisted mainly of Oak on the better soil, Beech on limy and upland tracts, and Scots Pine on the higher hills and on slopes having only inferior qualities of soil. If we look at the continent of Europe, we find the Scots Pine the chief tree over the bulk of the sandy stretches comprised within the North German plain, and the Spruce asserting itself throughout Scandinavia and north-western Russia and on the humid mountainous tracts of central Europe, pre-eminently on the Harz Mountains; a covering of Beech clothes the lower hills of central and north-western Germany, forming the Deister, Solling, and Thuringian forests; the Silver Fir occupies similar situations in south-western Germany and France, in the Black Forest, and the Jura mountains; whilst Austria has large tracts wooded chiefly with Black Pine and Larch, and Russia can show its well-defined areas on which Pines, Firs, Hornbeam, Birch, Alder, and Aspen are the dominant species. Speaking generally, the ruling species of trees throughout the forests of central Europe, as they now exist, may be said to be Scots Pine, Spruce, Silver Fir, and Beech in the first degree, followed by other Pines, Larch, Oak, Alder, and Birch. These are the species which form the bulk of the individuals in woodland crops; whilst Ash, Elm, Maple, Sycamore, Poplars, Willows, and exotics like Weymouth Pine and Douglas Fir, are only to be found in smaller numbers, or in situations specially suited for their growth. It is, however, rarely that any of these species are now to be found forming **pure forests** in countries where attention has been paid to the most recent developments of scientific sylviculture ; for the tendency is now rather towards the formation and maintenance of **mixed woods** as combining important sylvicultural and financial advantages. But, if entirely left to themselves, certain species of forest-trees will ultimately suppress and eliminate most other kinds over tracts of soil better suited to the former than to the latter. Evelyn was to a certain extent aware of this more than two hundred years ago, when he remarked, in one of his letters published in Aubrey's *Surrey*, that :—

'Where goodly Oak grew, and were cut down by my grandfather almost a hundred years since, is now altogether Beech ; and where my brother has extirpated the Beech there rises Birch.'

What might just be expected happens. The shade-bearing and densely-shading species—Beech, Spruce, Silver Fir, and in a less degree Hornbeam and Black Pines—assert themselves over large tracts ; whilst other kinds of trees of woodland growth are ousted, and occur merely as subordinate clumps, or groups, or patches, or as individuals scattered here and there over areas and situations unsuited for the predominating species, either on account of the nature of the soil or situation, or for some reason connected therewith (frost, want of moisture, shallowness of soil, &c.). There can, of course, never be rigid lines marking off the domain of the various species ; for there must always, in consequence of the various factors connected with the physical conditions of soil and situation, be belts of land where the advantage lies sometimes with one species and sometimes with the other. At the present time on the continent one has many opportunities of seeing how, in consequence of bad management and of servitudes, under which the peasants were entitled to rob the woodlands of their soil-protecting layer of dead foliage, the domain of the Oak and the Beech is being encroached on by the Pine and the Spruce, which are both more easily satisfied as to soil and situation ; whilst these

latter again have often to contend with the still more easily
satisfied Birch and Willow for the possession of their own
areas.   It might be argued against what is here stated, that
the Scots Pine asserts itself to the exclusion of shade-bearing
species on the vast plains south of the Baltic ; but this is solely
due to the fact that, for the poor sandy soil there, the Pine has
proved itself better suited and more accommodating than any
other species of tree, and is in fact merely another example of
the survival of the fittest.   Even on that sandy plain, near the
fertile lands along the riverine tracts, there is still to be found
a stately growth of Oaks, Ash, Elm, Maples, &c., able to hold
their own against any attempts at encroachment made by less
noble species of woodland trees ; whilst the Alder and Willow
coppices of those districts are famed.

From a study of the growth of trees in their natural habitats,
and of their development in varying situations into which they
may have been brought by artificial means, we are enabled to
classify and arrange them according to their sylvicultural
characteristics ; and a thorough knowledge of these is necessary
before we can hope to attain anything like the highest and
most successful results with regard to the formation, tending,
and reproduction or regeneration of woodland crops.   To give
only one instance of error made in regard to a valuable species
of tree, owing to want of scientific knowledge regarding its habit
and conditions of growth, take the case of the Larch, which
was introduced into England in 1629, into the Lowlands of
Scotland in 1725, and into the Highlands in 1727.   Of this tree,
27,000,000 plants are said to have been planted out in Scotland between 1738 and 1820 ; but it certainly has not fulfilled
all the anticipations that were then made as to its future value.
The Larch is essentially a tree of the mountains, where it
ascends higher than even the Spruce.   It is found on the
Bavarian Alps at from 3,000 to 6,000 ft. above sea-level in
excellent development and often forming pure forests ; lower
down it is naturally associated along with Spruce and Beech in

the zones more congenial to these genera.   The climate in which it thrives best is the Alpine climate, that is to say, one in which a long cold winter is succeeded by an exceedingly short spring, and followed by an equable, warm summer, which again, after a very short autumn, gives place to the long period of winter rest and defoliation.   Removed far from its true Alpine home, and subjected to climatic conditions essentially differing therefrom, can it be wondered at that its development has been, on the whole, disappointing?   In place of having the good straight bole, which is characteristic of its growth where indigenous, it tends to curved, sabre-like form; the stem and branches are prone to favour the development of hanging lichens (*Usnea barbata*, &c.); the foliage is damaged by cater-pillars of the mining-moth (*Coleophora laricella*) and of another Tortrix (*Grapholitha pinicolana*); whilst the boles are specially liable, at from about ten to twenty-five years of age, and more especially after bad attacks of these insects, to become infected with the serious cankerous, fungoid disease due to *Peziza Willkommii*.   This has often necessitated a clearance at the age of forty to fifty years, or long before the crops would, under more favourable circumstances, have been anything like tech-nically or financially mature.   The investigations of Professor R. Hartig of Munich have shown that the fungus which occa-sions this disease is indigenous to the Alpine home of the Larch, and sometimes occasions canker on stems there; but it has followed the tree to central and Northern Germany, and to England and Scotland, where, owing to climatic condi-tions and to its less vigorous growth, the stems are more liable to become infected, whilst the ravages committed are at the same time more serious.   His own words are as follows :—

‘ The enormous distribution which the Larch fungus, *Peziza Willkommii* has obtained on the German plains, is almost solely explained by the rich development of fully matured fruits and spores in the damp, and more especially in the stagnating, atmosphere of close-canopied forests in low-lying localities; whilst with the free play of aerial currents in Alpine tracts the fruits almost always dry up before they ripen[1].'

The crookedness of the stem may be due either to want of depth of soil interfering with the normal development of the strong tap-root, or else to the growth of the tree being stimulated beyond normal and healthy limits, owing to the much longer period of activity necessitated by our climate, in which spring and autumn are gradual and prolonged stages between the periods of foliation and defoliation.

[1] *Lehrbuch der Baumkrankheiten*, 2nd edit., 1889, p. 45.

CHAPTER III

THE CHIEF SYLVICULTURAL CHARACTERISTICS OF OUR WOODLAND TREES

THE differences in sylvicultural characteristics exhibited by the various species of our forest trees may be conveniently classified with reference to the following matters : —

    I. Climatic requirements.

    II. Requirements as to soil and situation.

    III. Capacity for bearing shade.

    IV. Normal shape of the stem and crown.

    V. Increment or rate of growth in height, girth, and cubic contents.

    VI. Reproductive capacity.

    VII. Attainment of economic maturity and normal duration of healthy growth.

When John Evelyn, as a Fellow of the Royal Society, read his *Sylva*, on October 15, 1662, we were quite abreast of continental knowledge relative to woodcraft; but since then we have lagged behind nearly every other country, whilst Germany has acquired by far the largest and richest stores of natural knowledge with respect to the science of Forestry. Yet, even in Germany, it is only within the last ten or twelve years that Professor Karl Gayer's observations and classifications have found general acceptance as the fundamental principles upon which the modern practice of Sylviculture—that branch of Forestry which treats of the formation, tending, and reproduction of woodland crops—has a firm, logical, and scientific basis. The first edition of his great work on Sylvi-culture was issued in 1880, and the final expressions of his matured conclusions, in the third edition of 1889, are the source from which most of the following data have been drawn [1].

I. Differences of Forest Trees as to Climatic Requirements.

The natural causes influencing the geographical distribution of the forest trees of Europe are partly climatic, and partly dependent on the physical conditions of soil and situation. The energetic growth and development of any particular species can only take place in localities that have such period of vegetation, and such intensity of summer warmth, as to enable the processes of assimilation and of the formation of ligneous tissue to proceed satisfactorily. For the least exacting species in this respect a three-monthly period of vital activity and a mean summer temperature of 54° to 57° Fahr. are the requisite minima [2]; whilst on the other hand the minima of the cold during the winter months also fixes their limits, so that large tracts of arctic Russia and Scandinavia are destitute of forest growth. The polar or northern limits of some of the more common forest trees are as follows :—

Species.	Western Europe (Scandinavia & Germany).	Eastern Europe (Russia).
Scots Pine . . .	68–70°	64°
Spruce	67–71°	54°
Oak (*Q. pedunc.*) .	63°	63°
Beech	60°	50–52°
Silver Fir . . .	49–52°	50°

And, in a similar manner, the reduction of the temperature

<hr>

[1] *Waldbau*, 3rd edit., 1889, pp. 18–49.
[2] Weber, *Die Aufgaben der Forstwirthschaft*, in Lorey's *Handbuch der Forstwissenschaft*, 1886, vol. i. p. 13.

in atmosphere and soil, with increasing height above sea-level, also causes a more or less irregular and inconstant distribution of species according to vertical zones in the mountains of central Europe, although in this respect the variations due to soil, situation and aspect are so great that it would perhaps be misleading to attempt any classification. It must, however, be remarked concerning the Scots Pine and the Spruce that their limits towards the north, and towards Alpine heights, are not so much due to the actual intensity of cold during winter, as to the fact that transpiration by the leaves is stimulated on bright sunny days in winter at a time when the frost-bound soil can yield no supplies of moisture to replace what is evaporated through the leaves [1]; hence the natural consequence is that the foliage grows yellow, the tree turns sickly, and death ensues. These results can most frequently be observed after long, dry, hard winters with frequent sunny days.

The southern limits of the forest trees of northern Europe are mainly determined by the quantity and the regularity of the rainfall during summer. As the results of careful experiments conducted in Austria, von Hönel [2] found that, computed by the unit of weight of their dry foliage, among broad-leaved trees, Ash and Birch require the largest supplies of water for transpiration, then Beech and Hornbeam, then Elms, and finally Maples and Oaks. Among conifers, the order was Spruce, Scots Pine, Silver Fir, Black Pine. He also found that, taken as a whole class, the broad-leaved species consumed on the average about ten times as much water as the conifers, and that, owing to the light foliage of the Pine, this species required very much less soil-moisture than Spruce or Silver Fir. But although the actual total of average annual rainfall may be considerably greater than what is required for transpiration through the foliage during the active period of

---

[1] R. Hartig, *Lehrbuch der Baumkrankheiten*, 2nd edit., 1889, pp. 104, 261.

[2] Compare, however, the more recent data set forth in Chapter IV.

vegetation (and, even for Beech, a summer rainfall of twelve inches would suffice according to Ramann [1]), yet when rain does not fall on the average once every 6 to 8 days throughout the summer months, the woodland crops are, in the dry climate of central Europe, exposed to more or less danger of being killed off by being unable to obtain from the soil the constant supplies of moisture requisite for the maintenance of healthy and vigorous growth by means of normal transpiration; or in other words, they are liable to suffer from drought and to die off.

So far as the climate of Great Britain is concerned, it is, both as regards temperature and atmospheric precipitations, suited to all the species of woodland trees that occur in central Europe; but the Mountain Ash, the Wych Elm, the Scots Pine and the Birch are more characteristic of, and grow better in, Scotland than in England, where again the Beech, the pedunculate Oak, the English Elm, and the Silver Fir thrive better than in the northern half of the main island. Pine, Spruce, Larch, Silver Fir, sessile Oak, and Sycamore, are naturally the inhabitants of upland tracts and mountain ranges, whilst Ash, Alder, pedunculate Oak, Aspen, Elm, and Willow belong rather to low-lying localities; Beech, Hornbeam, Maple, and Birch seem capable of thriving equally well on plains or on hilly country. But, on the whole, the nature of the climate, soil, and situation mainly determines the quality of any given species of tree in any particular locality.

Very little is as yet known regarding the absolute amount of atmospheric warmth necessary for the most favourable development of the different kinds of woodland trees. But experience has shown that Elm and pedunculate Oak require most warmth; that Black Pines, Silver Fir, Beech, Weymouth Pine, sessile Oak and Scots Pine can thrive with less than Birch, Maple, Sycamore, Ash, Alder, and Spruce; whilst Larch does with least.

So far as their development is apt to be prejudicially

[1] *Forstliche Bodenkunde und Standortslehre*, 1893, p. 311.

influenced by frost, Ash and Beech are most liable to injury; Oak, Silver Fir, Maple, Sycamore, Spruce, Douglas Fir, and Alder are somewhat liable; whilst Hornbeam, Elm, Birch, Larch, Aspen, and Pines are the most hardy species. It is not, however, the winter's cold that is injurious to these species, but the late frosts in spring, when the flush of young leaves and shoots is still tender, and the early frosts in autumn before the summer shoots have had time to harden sufficiently so as to resist the effects of the cold.

The relative humidity of the atmosphere is also a climatic factor exerting very considerable influence on the vigour and general development of the various kinds of woodland crops. Experience has shown that, so far as any difference can be made in a generally damp climate like that of our islands, the species thriving best in localities with great relative atmospheric humidity are Willow, Poplar, Alder, Maple, Sycamore, and Ash; Silver Fir, Scots Pine, Beech, Aspen, Birch, and Douglas Fir do best with a moderately damp climate; whilst Oak, Elm, Black Pines, and Larch naturally prefer a rather dry atmosphere. It may be of interest to note that the Scots Pine, which thrives so well in our moist insular climate, does not thrive in Denmark, and is essentially a tree of dry localities in Germany; whilst Spruce, which thrives best in humid localities throughout central Europe, does not appear to grow anything like so well in Britain.

II. Requirements as to Soil and Situation.

The aspect or exposure of any particular locality tends to exert an influence in the same direction as vertical elevation, though on a much smaller scale; for, as the mean annual temperature of slopes facing from S.E. to S.W. is higher than that of hill-sides with N.E. to N.W. aspect, this naturally affects the thriving of woodland crops. Near their limits of vertical distribution, however, all trees ascend the mountains to a greater height on the southern than on the northern sides, in consequence of their requirements as to warmth.

The root-systems of the various species of forest trees vary greatly with regard to shape and to the depth to which they reach ; but even shallow-rooting kinds derive advantage when the soil over which they grow is deep, owing to the greater supplies of nutriment within easy reach of their roots. Some are characterized either by a strong tap-root, or else by a deep heart-shaped root with stout ramifications, as in the Oak, Elm, Pine, Silver Fir, Maple, Sycamore, Ash, and Larch ; others have no pronounced tap-root, but develop strong side-roots penetrating into the soil for a moderate depth, as in Beech, Hornbeam, Aspen, and Birch ; others again, like the Alder, throw out strong side-roots, whence strands are sent down into the soil; whilst some are unmistakably shallow-rooting, as in the case of the Spruce. Spruce, Aspen, and Birch require least depth of soil, and Oak and Larch greatest. Of the other trees, Scots Pine, Silver Fir, and Douglas Fir need deeper soil than Austrian and Weymouth Pines ; whilst Beech, Hornbeam, and Alder, although by no means shallow-rooted, can do with a less depth than Elm, Maple, Sycamore, and Ash.

In referring to von Hönel's experiments regarding the transpiratory power of the various species of trees per unit of weight of foliage, it will be remarked (Chapter IV) that his researches give us no reliable data for estimating the absolute quantities of soil-moisture necessary for woodland crops so as to supply their normal requirements throughout the annual period of active vegetation.

Practical experience, however, shows that Alder, Ash, Willow, Poplars, Maple, Sycamore, and Elm, then Larch, Weymouth Pine and Spruce, prefer soils with a considerable degree of moisture ; whilst pedunculate Oak, Hornbeam, Birch, and Aspen, and in a less degree Silver and Douglas Firs, also make greater demands than Beech and sessile Oak, or Scots, Corsican, and Austrian Pines, which are the three species that can best accommodate themselves to dry soils (the former on those of a sandy, the latter on those of a limy nature).

On the whole, the conifers are more moderate in their demands on soil-moisture than the broad-leaved genera of trees. But the limits within which the different trees can thrive vary greatly according to the species; or in other words, the various kinds of woodland trees exhibit great differences as to their accommodative power with regard to the degree of moisture contained in the soil.

Although trees derive the greater part of their nutriment from the carbonic acid which they absorb from the atmosphere, still they are also dependent on the soil for supplies of mineral nutrients; and these can only be taken up when held in solution by the soil-moisture. Whether the soil be loose or binding is also a matter of very considerable importance to the thriving of the various woodland trees. As a rule the broad-leaved species do better than conifers on the stiffer classes of land, although soils of merely average tenacity are on the whole most suitable for all kind of trees, owing to their better endowment as regards physical properties. The chief constituents of soils are clay, lime, and sand; and as clay yields the most valued nutrients for plants, the qualities of soils are often determinable to some extent by the quantity of clay found in them. The effects of the mineral constituency of soils are, however, very greatly modified by the physical properties inherent in them, and these become prejudiced whenever any one of those three most important constituents occurs in undue excess. Clay soils are tenacious and interfere with the movement of soil-moisture; sands are too porous, and limes are too easily heated; hence on the average, the mixed classes of loamy soils are in general most fertile. The absolute demands of the various species of trees on the mineral strength or fertility of soils is not yet fully known, although they can to a certain extent be estimated from the weight and analysis of the ash left after reducing the different kinds of timber by combustion.

On this point, however, sylvicultural experience shows that Elm, Maple, Sycamore, and Ash, make the greatest demands; and that Oak, Beech,

Silver Fir, and Willow are more exacting than Larch, Hornbeam, Alder, Spruce, and Aspen ; whilst Poplars, Pines, and Birch can thrive fairly well on indifferent classes of soil.

Here again it may be noted that, on the whole, the broad-leaved species of trees are more exacting than the conifers. As a broad generalization it may be asserted that the power of any species to accommodate itself to soils and situations not naturally suited to its normal requirements varies in the inverse ratio of the demands it makes as to mineral strength ; for it is greatest in Scots Pine and Birch, and least in Elms, Maple, Sycamore, and Ash.

It would be wrong to state broadly that a clayey, limy, sandy, or loamy soil is, or is not, suited for any given species of tree ; because so very much depends on other considerations, and particularly on the amount of humus present in it, owing to the very favourable influence which this exerts on the physical properties of the soil.   Gustav Heyer even went so far as to maintain [1] that almost any soil was capable of producing any given kind of timber, provided that it contained the requisite amount of moisture.

Ney [2] thus summarizes the best practical opinions that have yet been expressed on this subject :—

'As regards the chemical composition of the soil, even slightly sour marshy soils are unfavourable to all species of trees except Alder, Birch, and Spruce ; whilst sour soils, liable to dry up at certain seasons, are unsuited to all except Birch, Spruce, Scots and Weymouth Pines. Only these last-named species thrive on pure peat, and not even the Spruce when it is dry.   Ash, Maple, Sycamore, and Elm require a moderate quantity of lime in the soil, and Beech, Hornbeam, Oak, as also Larch and Austrian Pine, thrive best on soils that have at least some lime in their composition.   The hardwoods—Oak, Ash, Maple, Sycamore, Elm, Chestnut, Beech, and Hornbeam—also appear to demand the presence of a considerable quantity of potash ; whilst, on the other hand, Spruce, Silver Fir, and especially Scots Pine and Birch, thrive on soils rich neither in lime nor potash.'

[1] *Forstliche Bodenkunde und Klimatologie*, 1856, p. 488.
[2] *Lehre vom Waldbau*, 1885, p. 64.

The influence of the sum total of the factors dependent on soil and situation find their expression in the amount of timber produced per acre, and in the quality of the timber, since this latter must determine its adaptability for technical purposes. Practically speaking, the best and the most convenient general standard for estimating the quality of timber is to be found in the specific weight, as, *for one and the same kind of wood*, the heavier specimen is the more durable[1], owing to the thick deposits of ligneous substance in the cells. This does not, however, apply to any comparison of different species; for the light wood of conifers is more durable than heavy Beech timber. High specific gravity, length, straightness, and full-woodedness of bole, with freedom from knots and branches, are the best indications that soil and situation are eminently suitable to the woodland crops covering them.

III. Capacity for bearing Shade.

All trees require a certain amount of light in order to carry out the work of assimilation of the carbonic acid which forms so important a part of their nutriment. Hence the want of a due intensity of light exerts an injurious influence; otherwise there is no reason why the density of the foliage in the interior of the crowns of trees should not be as great as it is near the circumference. It is true that excess of light may tend to paralyse the action of chlorophyll; but, so far as the forest trees of Britain are concerned, they are exposed to no danger from this cause in our climate; they will thrive all the more vigorously the more light they receive the benefit of. But, with regard to the measure of light necessary for the performance of the assimilative functions by the foliage—or, in other words, to the capacity for bearing shade—marked differences occur among the various species of trees; and though affected by soil

---

[1] Gayer, *Die Forstbenutzung*, 7th edit., 1888, p. 66.

and situation, these differences are, especially in woodlands, sufficiently distinct to admit of general classification into shade-bearing trees, light-demanding trees, and trees that occupy an intermediate position. The different species of our forest trees may accordingly be classified as :—

Light-demanding trees:
>   Larch ; Birch.
>   Scots Pine; Aspen, Poplar, Willow.
>   Oak, Ash, Elm ; Douglas Fir.

Trees less impatient of shade:
>   Alder, Maple, Sycamore; Weymouth, Corsican, and Austrian Pines,
>   and Menzies Spruce.

Shade-bearing trees:
>   Spruce; Hornbeam.
>   Beech ; Silver and Nordmann's Firs.

The capacity of the various species for bearing shade under any given circumstances, or in other words, their demands for light, may be gauged by the general density of the foliage of the crown and the capacity of overshadowed twigs to retain life. But, so far as regards the absolute quantity of light requisite for the performance of the assimilative functions of any species, we know as little as about the necessary amount of warmth. For general sylvicultural purposes, Larch, Birch, Scots Pine, Oak, Ash, Aspen, and Willow, may be considered the principal light-demanding species of trees, and Silver Fir, Beech, Spruce and Hornbeam the chief shade-bearers ; whilst the attitude of the other woodland trees in this very important matter is mainly dependent on the conditions of the soil as to depth, moisture, and general quality.

The length of the annual period of vegetation also makes itself felt to a certain degree ; for in the far north, or at high elevations, where this is shorter than in lower latitudes or levels, the demands for light become greater than obtain under the opposite conditions. The same may also be said of cool northern exposures and misty regions, as compared with warmer southern aspects and localities where the crowns can revel in the undisturbed possession of genial sunlight and warmth.

The relative powers of the various species of trees to bear shade under any given conditions as to soil and situation are of great, and indeed vital, importance in practical sylviculture. For on these are mainly dependent the measures that may seem prudent with regard to the selection of species of trees for woodland crops, the extent to which admixture of species may be advisable, the best method of treatment and tending to be accorded to the woods, the most advantageous time for harvesting the mature crops and reproducing them, &c. In fact, without a due consideration of their individual characteristics in this respect, it is impossible to accord rational scientific and practical treatment to woodland crops.

IV. Normal Shape and Development of the Stem and Crown.

The essential characteristic of a tree as compared with a shrub is its habit of developing a bole or stem before beginning to ramify and form a crown of foliage-bearing branches. Some kinds of forest trees have a greater tendency to ramification than others, as, for instance, the broad-leaved trees generally in comparison with conifers; but the individual tendencies of each kind with regard to the length and straightness of the bole are often to a very considerable extent dependent on (1) *the growing-space allotted to it*, (2) *the age of the crop*, (3) *the nature and quality of the soil and situation.*

Douglas Fir, Spruces, Silver Firs, Larch, and Weymouth Pine, then Scots Pine, sessile Oak, and Alder have naturally a strong tendency towards the formation of a good bole or stem ; whilst Hornbeam, pedunculate Oak, and Willow are prone to branching growth.

The other species of woodland trees occupy an intermediate position between these two classes. But the natural tendency to assume their normal shape has only free scope when the individual trees are growing in full exposure to light and air. Those in which the leading-shoot develops more vigorously than the side-shoots then assume a conical form; whilst others, whose side-shoots compete on anything like equal terms with the axial shoot, remain short and stunted, and have low wide-spreading crowns. The true characteristics of any tree are best observable when it is grown in the open; but, when the **growing-space** allotted to each tree is limited, these natural tendencies are checked, and a struggle upwards for individual exposure to light and air ensues, which transforms the whole vital energy into activity of growth upwards. This impulse is greatest in light-demanding trees, and, within reasonable limits, may be said to vary inversely to the growing-space allowed. Even when grown in the open, Spruces, Douglas Fir, and Silver Firs, Larch, and Weymouth Pine retain a distinctly noticeable central axis, which is yet traceable, though less clearly, in Scots, Corsican, and Austrian Pines, Alder, Beech, and sessile Oak, and to a still less degree also in Ash, Maple, Sycamore, and Elm; whilst on the other hand the pedunculate Oak and Hornbeam have a marked tendency to ramification and to the formation of a diffuse crown at no great height above the ground. In woodland crops of normal density the leafy crowns of Larch, Pines, Oak, Birch and Aspen reach only a little way down the stem; those of Spruce, Douglas Fir, Beech and Hornbeam descend for about one-third of the bole; whilst in Silver Fir they often extend almost half-way down. The effect of the concentration of the crown of leafy foliage towards the summits of the stems ensures the conservation of the assimilated nourishment for the formation of long, straight, full-wooded stems of the highest possible technical and financial value, in place of this being dissipated over a large branch-system and a stunted stem. That such timber must be more full-wooded, i.e. that the relative proportion between the upper girth or diameter and the lower should show less difference, is easily intelligible, owing to the larger supplies of assimilated nourishment offered under the above circumstances to the upper portion of the bole than to the under part.

In the earlier stages of development **the age of the crop** is of little practical importance. All species of trees, when grown in woods of normal density, assume a conical or spindle-shaped crown till they outgrow the pole stage and become tree-forest; but, having passed through the most active period of their growth in height, they then begin to exhibit their natural tendencies more distinctly. Thus Larch, Spruces, Douglas Fir and Silver Firs continue to develop more vigorously upwards than side-wards; Pines, Elm, Beech, Maple, Sycamore, Birch, sessile Oak, Ash, and Alder begin to assume an oval crown; whilst English Oak and Hornbeam get rounded off with a broad, obovate crown. As they approach the limits of age of healthy vigorous growth, all trees, except Larch, Spruce, and Douglas Fir, assume a blunted or rounded off growth near the summit, due to the growth in height declining sooner than the lateral expansion.

So far as **soil and situation** affect the shape and habit of growth, fresh, fertile loams stimulate to full-foliaged coronal development and large girth of bole, though somewhat at the expense of its length. Deep, fresh, light sandy soils favour length of stem; but, as the crowns are thinner, and the branch-development is sparser, the girth consequently remains less. Shallow and rocky soils affect the stem-development very prejudicially, and cause branching growth and excessive root-systems spreading far around in search of moisture and nutri-ment. Great elevation above sea-level, and raw, cold exposures, militate against the formation of good stems. In Scotland, at 2,500 ft., the Scots Pine is more like a shrub than a tree.

The conclusion deducible from the above facts is that wood-land crops formed of shade-bearing genera able to thrive with a small growing-space, i.e. Beech and Hornbeam, but more particularly Spruces, Silver and Douglas Firs, are able to main-tain close canopy better and longer than woods formed chiefly

of Larch, Pines, Ash, Maple, and Sycamore. And when, as
in the case of Oak, Birch, and Chestnut, a tendency to branching growth is combined with strong demands for light, the
leaf-canopy is apt to be interrupted early and to an injurious
extent; and the natural impulse in this respect is all the
greater, the less favourable the soil and situation may be for
the particular species of tree in question.

V. Increment, or Rate of Growth in Height, Girth, and Cubic Contents.

Growth in height varies, both in energy and duration, according to the species and the age of the tree, the soil and situation,
and the methods of formation and treatment of the woods. It
is most active in Douglas, Spruce, and Silver Firs, Larch, Scots
and Weymouth Pines, which can attain an average height of from
110–140 ft.—the Douglas Fir grows to over 300 ft. in height,
and 18 ft. in girth, in its home in North America; Oak, Ash,
Beech, Maple, Sycamore, and less frequently Elm, Poplar, and
Birch reach a height of 100–130 ft.; whilst Black Pines, Hornbeam, Alder, and Willow are seldom over 80–100 ft. high when
mature.   These data are only for crops of normal density.

But the maximum heights usually attained by the various
species of trees are reached in different periods of time. This
fact, combined with their varying demands as to light, is of
immense importance from the practical sylvicultural point of
view; as these are two of the main points to be considered
(when once the protection of the productive capacity of the
soil has been duly safeguarded) with regard to the formation
and tending of mixed woods.

Birch and Larch, above all others, then Poplar, Alder, Ash,
Maple, Sycamore, Elm, Willow, Weymouth and Scots Pines,
and Douglas Fir, are the trees that shoot up most rapidly
during their youthful period of growth; whilst Oak, Beech,
Hornbeam, and Black Pines may be classed as of moderately

rapid growth, and Spruces and Silver Firs as relatively slow in initial development.

These results show that, in a general way, the rate of growth in height during the earlier stages of tree-life coincides with the demands made by the species on light and growing-space. Some species, like Douglas Fir, Larch, and on good soils Pines and Birch, maintain the advantage gained; but others, like Ash, Maple, Sycamore, and Aspen, are caught up in growth and overtopped by Spruce, Silver Fir, Beech, and Oak, whose vigour increases on their attaining the pole-forest stage of development. The duration of active growth in height in any given species depends to a great extent on the quality of the soil and situation; but, *coeteris paribus*, it is maintained longer by Larch and Firs than by Pines among conifers, and longer by the sessile Oak, Elm, and Beech than other species among broad-leaved trees. The average heights (in feet, and approximate only) of some of our more common forest trees, on soils of merely average quality for each different kind of tree, are about the following in well-managed woods of normal density :—

Age of Crop. Years.	Beech.	Silver Fir.	Spruce.	Scots Pine.
20	$10\frac{1}{2}$	10	11	$17\frac{1}{2}$
40	34	35	33	39
60	$52\frac{1}{2}$	$58\frac{1}{2}$	57	55
80	67	72	$75\frac{1}{2}$	65
100	76	$81\frac{1}{2}$	88	75
120	83	85	$95\frac{1}{2}$	80

It is at once perceptible from these figures how careful the tending will require to be in mixed crops between the fortieth to sixtieth years, and even earlier on soils of inferior quality; for wherever light-demanding species are overtaken by shade-bearing species they must be killed off in the struggle for supremacy, unless favoured to a greater or less extent during the special operations of tending.

Growth in girth is in all species of forest trees practically proportional to the growth in height, as they are each the expression of the vigour of the individual tree. In most of the light-demanding species of trees this begins early, often attains its maximum vigour between the twentieth to thirtieth year, maintains itself till about the fiftieth to sixtieth year, and then gradually declines. It begins later with Oaks and shade-bearing species, but is often vigorously maintained till the seventieth to ninetieth year before beginning to sink gradually. In crowded woods the development in girth is more prejudiced than the energy of growth in height ; but whenever the growing-space allowed is more than sufficient for the normal requirements of any given species, growth in girth takes place at the expense of growth in length. When trees still in vigorous growth are heavily thinned after being accustomed to but a limited growing-space, they rapidly thicken in the bole in consequence of the greater exposure to light and air, and of the increase in foliage and assimilative power due thereto. The largest girths are attainable by Douglas and Silver Firs, Spruces, Weymouth and Scots Pines among conifers, and by Oak, Elm, Beech, and Black Poplar among broad-leaved species of trees.

Growth in cubic contents, or total increment, is the final expression of the resultants of growth in height and growth in girth, and is a convenient measure of the general energy of growth. Owing to the reduction in the number of individual stems per acre, consequent on increased demands for growing-space, increment culminates earlier in light-demanding than in shade-bearing species. So far as whole crops are concerned, Gayer classifies their energy in total increment as follows :—

> Spruce and Silver Fir woods.
> Larch, Weymouth and Scots Pine woods.
> Beech woods.
> Oak, Ash, and Hornbeam woods.
> Birch woods.

But he specially characterizes the energy of the Firs as nearly twice as great, and that of the Pines as about half as great again as that of Beech, which is the most energetic of the broad-leaved species. It will be seen, from what has previously been said, that the productive energy of the individual tree may be widely different from the productive energy per acre of a pure forest of that species; and it will therefore readily be intelligible how advantageous—alike from the material, the financial, and the sylvicultural points of view—the judicious formation of mixed crops may be as compared with pure forests, more especially if formed of the light-demanding species.

VI. Differences with regard to Reproductive Capacity.

Trees may reproduce themselves either by means of seed, or else by means of shoots from the stool, or through suckers thrown up from the roots. Reproduction from seed is the normal process of regeneration of all species that are indigenous, or that have thoroughly naturalized themselves in our climate; whilst the formation of stoles or root-suckers is principally confined to English Elm, and to exotic Poplars and Willows, which cannot be relied on to form seed of anything like average germinable capacity; of indigenous trees, Aspen and Willow are the only species endowed with this quality to any great extent. The formation of shoots from the stool is not so much a regenerative measure as rather an effort at recuperation and replacement of the stem or bole when the ascending axis has been removed. It is solely due to the vital activity of the root-system, which, until late in life, remains active after the stem and crown have been removed, and which endeavours, by a new flush of shoots and foliage from the adventitious or dormant buds at the neck of the stump, to obtain the means of assimilating the food that it continues to absorb from the soil. This same tendency is seen in pollarding as well as in coppicing, but in another form and a more limited degree.

The formation of seed depends on the supplies of starchy and nitrogenous reserves stored up in the tree. These vary according to its age, the soil and situation, the amount of exposure to light, and the warmth of the summer period of vegetation. In good seed-years Beech, Oak, Spruce, Pine, Birch, Hornbeam, Elm, Aspen, Alder, and Willow produce larger quantities of seed than Ash, Maple, Sycamore, Silver Fir, and Larch. But, classifying them as to the total quantities of seed produced during long periods, it may be said that Birch, Aspen, and Willow are most prolific ; next to them come the Pines, Spruces and Douglas Firs, Elm, Hornbeam, and Alder ; behind these again rank the Oak, Maple, Sycamore, Silver Fir, Larch, and Ash ; whilst last of all in this respect comes the Beech.

From this it cannot fail to be noted that in general the species with small seeds are more prolific than those with large and heavy fruits. As these light, and very often winged, seeds are more easily wafted and borne far away by winds, it must be admitted that species of trees like Birch, Aspen, Pine, and Spruce have better natural reproductive power than Oak, Beech, Silver Fir, and the like. It is worthy of note, too, that the former are species which make less demands on soil fertility than the latter, and are at the same time endowed with greater accommodative powers as to soil and situation ; for these are matters almost sure to be closely connected with the power of accumulating reserves of starchy and nitrogenous substances utilizable for the formation of flowering-buds.

The different kinds of seeds vary, however, greatly in their germinative power. This is lowest in Birch, Alder, Elm, and Larch. It amounts to about 50 % in Ash, Hornbeam, Maple, Sycamore, Weymouth Pine and Silver Fir ; whilst it is somewhat greater in Oak, Beech, Spruce, and Scots, Corsican, and Austrian Pines.

The seed of most species of trees germinates in the spring after it has been shed ; but that of Birch, Elm, Aspen, and Willow sprouts during the spring in which it falls, and that of

Ash and Hornbeam germinates only in the second spring after its fall, if the seed has been stored before being sown.

Seed-production is greatest when the main growth in height is completed, and the natural tendency to coronal development makes itself apparent in increased demand for light and growing-space, i.e. when the vital energy is at its maximum. Warmth of situation, fertility of soil, and increase in growing-space, all stimulate to production of seed, owing to the larger supplies of nutrients, and the better assimilative opportunities then available for each individual tree.

The formation of stool-shoots taking place in coppice-growth is essentially an expression of recuperative power. It is inherent in the broad-leaved species to a very much greater degree than in conifers, among which, indeed, with the exception of the Larch and the three-needled species of Pines, it is almost practically non-existent. This power of replacing the stem and crown is greatest during the younger stages of growth, but is always to a great degree dependent on the quality of the soil and the degree of exposure to light. Oak and Hornbeam retain their recuperative power in this respect longest (to about 85 years of age), whilst stools of Beech and Birch, after being coppiced several times, lose their reproductive vigour.

The formation of stoles or suckers appears to be the only means of utilizing their surplus reserves of starchy materials with which exotic species of trees (English Elm, Lime, Chestnut, and most Poplars and Willows) are endowed, seeing that from climatic causes they are unable to utilize them in the same way as indigenous trees, or the other exotic species that are less exacting with regard to warmth, that is to say, in the normal production of seed of average germinative capacity.

Stool-shoots are more often produced than suckers by Oak, Hornbeam, Beech, Elm, Alder, Ash, Maple, Sycamore, Willow, and Birch; whilst stoles are more frequent than stool-shoots from Aspen, non-indigenous Willows and Poplars, and White Alder. But Willow, Chestnut and Elm at the same time throw out a very fair proportion of stoles, which may be cut away and transplanted like seedling growth.

VII. Differences as to Attainment of Maturity and normal Duration of healthy Growth.

The ages to which the various kinds of forest trees may be grown in a healthy condition without showing visible signs of senile decay exhibit great differences. Thus we find Oak and Scots Elm attaining 500 years and more, English Elm, Silver Fir and Beech 300–400 years, Ash, Maple, Sycamore, Spruce, Larch, Scots Pine and Hornbeam about 200 years, whilst Aspen, Birch, Alder and Willow seldom attain over 100 years. Indeed, many historical trees are known to be very much older than any of these limits ; but even the above ages are far in excess of any rotations that could possibly be maintained in woodlands worked on sound economic sylvicultural and financial principles.

For the attainment of healthy old age the essential conditions are (1) that the opportunities of growth must be such as to permit of the normal development of all the organs of nourishment, and (2) that the soil and situation must continue to supply all demands as to warmth, moisture, &c., made by the individual species. But very frequently the economic requirements of sylviculture are incompatible with the demands as to growing-space, &c., that are involved in the above. And the capacity of different genera for regaining normal vigour, after being first grown for a long time in close canopy and later on allowed a larger measure of light and air, varies considerably ; it is greatest in Oak, Lime, Willow, Elm and Silver Fir, and least in Alder, Aspen, Maple, Sycamore, Beech, Hornbeam, and Spruce.

From the practical sylvicultural point of view, the following may be regarded as the ordinary average limits of age which it is advisable to allot to timber crops grown in high forest; but so much depends in each case on the concrete conditions as to climate, soil, and situation, that the figures must be regarded as rough generalizations only :—

Kind of Tree.	On soils of good quality.	On soils of inferior quality.
	Years.	*Years.*
1. Silver and Douglas Firs	100–120	60–80
2. Spruce	80–100	60–70
3. Scots Pine and Larch .	80–100	50–70
1. Oak	120–150 ; 180–200	90–120
2. Beech	90–120	80–100
3. Hornbeam, Elm, Maple, Sycamore, Ash . .	60–80	40–50
4. Willow, Birch, Aspen .	50–60	40–50

From these approximate data it will be at once apparent that when mixed crops are formed of several genera, these may attain their technical and financial maturity at different ages; hence it is the duty of the forester to utilize those which mature soonest, by extracting them at the time of their highest economic value.   This is done not only with a view to obtaining the maximum monetary returns, but also for the purpose of protecting, to the greatest possible extent, the productive capacity of the soil.   For the realization of both of these objects abundant opportunities are given during the process of **tending the timber-crops** throughout all stages of their development, to the consideration of which matter one cf the succeeding chapters will be devoted (see Chap. IX).

CHAPTER IV

THE NUTRITION AND FOOD-SUPPLIES OF WOODLAND CROPS

'Concerning the growth of plants a large amount of information has been amassed, but we are far from possessing even an approach to a knowledge of the laws which regulate this important subject.'—ROSCOE's *Elementary Chemistry*, 1888, page 413.

ALL plant life is governed by what is known in agricultural chemistry as the **Law of the Minimum**. According to this law the essential factor occurring *in minimo* regulates the total extent of production; whilst any given species of plant attains its finest growth and development in localities where all the essential factors are most favourably combined. These essential factors are partly of a physical, and partly of a chemical nature. To the former belong the action of warmth and of light, and to the latter carbonic acid, oxygen, and the water requisite for dissolving and holding in solution the mineral nutrients that can only be absorbed by the plant in the shape of soluble salts.

Temperature makes itself felt, throughout both the soil and the atmosphere, in calling into activity the process of vegetation in spring, and in continuing it during the summer and the autumn. Most of the trees indigenous throughout central and northern Europe begin their period of active vegetation in spring when the temperature rises to 6° or 8° C., although with exotics from the warmer south a higher temperature is requisite in order to stimulate them to activity.

This is a subject which has hitherto been little studied, as the following extract from Willkomm [1] may show :—

'Despite a mean winter temperature of + 17·5° R. (71·4° Fahr.) the Beech in Madeira has a period of rest for 149 days, during which it remains leafless !　This strange phenomenon is to be noted also in the Elm, Silver Poplar, Willow, and other broad-leaved trees distributed through the whole of Europe, as well as in the Apple, the Pear, and other fruit-trees over all the most southern parts of Europe.　They also lose their foliage in autumn, and only flush anew in spring, although during the whole of winter the temperature does not sink below + 5 to 7° R. (43¼° to 47¾° Fahr.).　And whilst in central and northern Europe the buds of the above-named trees open early in spring after warm winters, this is not the case in the south.　This phenomenon must be due to some law of nature which remains as yet unknown.'

And in another place [2] he remarks with regard to woodland trees that—

'The absolute minima during the period of vegetation are of far greater importance, with regard to the thriving of trees, than the minima of winter, and the average temperatures of the months and seasons.'

Indeed he goes so far as to say that [3]—

'It may be stated as a general law that *the polar or northern limit of any particular species of tree is determined by a certain definite isotherm of July, whilst its equatorial or southern limit is determined by a certain definite isotherm of January.*'

Seebohm's observations in Siberia agree with the first part of that statement.　For he observed that [4]—

'The limit of forest growth does not coincide with the isotherms of mean annual temperature, nor with the mean temperature for January, nearly so closely as it does with the mean temperature for July.'

When once active vegetation has been stimulated by increasing

---

[1] *Die Forstliche Flora von Deutschland und Oesterreich*, 2nd edit., 1887, p. 452.
[2] *Op. cit.*, p. 364.
[3] *Op. cit.*, p. 39.
[4] Presidential Address to Geographical Section of the British Association, 1893, in Royal Scot. Geog. Society's Magazine, October, 1893, p. 511.

warmth in spring, the energy of the vital process is enhanced as the temperature rises, until it may reach a point at which the plant is unable to carry out the vital operations in a normal manner.   In Britain, hot southern slopes with light sandy soil may often prove too warm for woodland trees like Spruce and Silver Fir, whilst cold land with a northern exposure may not supply the amount of warmth requisite for the best development of Oak.   In forestry, no measures can be taken to regulate the atmospheric temperature.   But by maintaining a close canopy of foliage, or by the underplanting of standards, much can be done to regulate the soil-temperature, and to retard the commencement of active vegetation.   This may even be delayed for about a fortnight or more, and may thus be made to commence only at a time when, the atmospheric temperature and the intensity of the light being greater, the process of assimilation may be more thorough as well as more energetic.

Light, which is closely allied to heat, is essential in order that, with the aid of the chlorophyll contained in the leaves, the decomposition of water and of carbonic acid may take place in the elaboration of substances to be used for structural purposes; whilst at the same time various other processes, such as transpiration, the deposition of matter on the cell-walls, and, in fact, the whole nutrition, are all likewise more or less dependent on the action of light.   The intensity of light necessary for the different species of trees, before they can carry out the process of assimilation, varies considerably, as may be seen from the different densities of their foliage; whilst the amount and intensity of light required by one and the same species of tree also varies according to the other factors of the soil and situation.   Thus, for example, under unfavourable circumstances of soil the foliage is less dense, and the leaves are in every way smaller, than if the same species of tree be growing on more favourable localities.   Leaves growing in full exposure to light are thicker, contain more chlorophyll in their parenchym, and have a much larger number of stomata than those growing in shade. The latter are sometimes only about one-third of the thickness of the former, and great differences in size may even be noticed in the buds throughout the winter. If young growth, that has hitherto stood in shade, be suddenly exposed to the full action of light, these thinner leaves are not always capable of carrying out the new and more energetic work thrust upon them ; hence they often fall into a sickly condition, which predisposes them to disease, and may even induce the death of the plant. Thus, in areas regenerated naturally, the clearance of the parent standards should be made gradually, although the period over which it must be extended varies considerably according to the kind of tree, and, for one and the same species, according to the concrete conditions of the soil and of the plants concerned. Broad-leaved deciduous trees can in general accommodate themselves comparatively easily to rapid transition from shade to light, owing to their leaves only lasting for one season ; but in conifers this accommodative power is necessarily more gradual, and is least in those genera that retain their foliage for a long period, like Spruces and Silver Firs. The poorer the soil of any locality, and more particularly the nearer it approaches the minimum with respect to the water-supply requisite for the normal transpiration of any species of woodland tree, the greater is the necessity for a free supply of light in order to carry out in as thorough a manner as possible the assimilative process with the limited amount of sap and of nutriment yielded by the soil. But the quantity of foliage decreases along with these latter supplies. Hence it becomes of all the more importance that the limited amount of nutriment should be elaborated in a most thorough manner ; and this is only possible with the comparatively free exposure of each individual leaf to light and air. Or, in other words, this naturally leads to the formation of a loose crown of foliage, instead of a dense leafy canopy being maintained.

Carbonic Acid or Carbon Dioxide (CO_2) becomes decomposed under the action of light by the chlorophyll contained in the parenchym of the leaves, and is assimilated, or prepared for being made of use for structural purposes, in different combinations of carbon [1].   As on the average 0·03 % by volume, and 0·05 % by weight, of the atmosphere consists of carbonic acid, there is never any want of this essential factor to the due thriving of any kind of woodland tree in any given locality.

Whilst there is no reason to deny to ordinary plants, with foliage containing chlorophyll, a certain capacity for absorbing nutriment directly from *humus* or organic matter in the soil, as is done by those which are not provided with chlorophyll, yet the thriving of woodland growth on soils poor in *humus* is of itself a proof that any direct absorption of carbon compounds by the root-system cannot be of essential importance to their existence.   The discovery, by Sadebeck and Frank about nine or ten years ago, of the mantle of fungoid tissue around the extreme points of the rootlets of cupuliferous trees (Oak, Beech, and Chestnut), and since also proved in the case of other woodland trees, coniferous as well as deciduous, has given rise to a theory that the introduction of organic matter is often carried on by means of the symbiosis of a fungus or *Mycorhiza* at the points of the suction-roots.   This *Mycorhiza* is supposed by Frank to act as a go-between, absorbing water, and nitrogenous and non-nitrogenous organic matter from the soil, and giving these to the rootlets, from which again the fungus receives other nourishment in exchange.   According to Ramann [2], however, the appearance of the *Mycorhiza* must be considered as already the indication of an abnormal or diseased condition, rather than

---

[1] The chemical process is probably as follows (see R. Hartig's *Anatomie und Physiologie der Pflanzen*, 1891, pp. 172 and 229): $C_2O + H_2O = CH_2O + 2O$ set free; $6(CH_2O) = C_6H_{12}O_6$ dextrin, glucose, or grape sugar, in which form the assimilated product is capable of being translated from the leaves to other parts of the organism.   When formed quicker than it can be transported, this loses one molecule of H_2O, and becomes $C_6H_{10}O_5$ or starch, which is secreted temporarily in the solid form and transformed again into grape sugar, when required, by resuming the molecule of H_2O.

[2] Ramann, *Forstliche Bodenkunde und Standortslehre*, 1893, p. 302.

a genuine symbiosis bringing mutual advantages both to the fungus and the tree.

Oxygen (O), requisite for the process of respiration, is, like carbonic acid, always present in sufficient quantity in the atmosphere, which is composed of 20·93 % oxygen and 79·04 % hydrogen by volume, or 23·28 % oxygen and 76·67 % hydrogen by weight. Unless, however, the aeration of the soil is also favourable, it may happen that the root-system has difficulty in finding a sufficiency of oxygen for the supply of its requirements.

Ebermayer found that under Beech the soil was always much poorer in carbonic acid than under crops of Spruce and on fallow land, a fact which he ascribed to the better **aeration of the soil** by the numerous and deep-going roots of the Beech. Ramann agrees with this view and adds [1] :—

'The "soil-improving" action of the Beech is probably mainly ascribable to its thorough aeration of the soil. As a large percentage of carbonic acid in soils rich in humus indicates a considerable reduction in the amount of oxygen in the air circulating throughout the soil, there is no reason why the former should be advantageous to plant life—on the contrary, the greater amount of carbonic acid may rather be taken to be a sign of deficient aeration and of deterioration in the soil.'

Prof. Ebermayer's own most recent utterance with regard to the soil-improving qualities of the Beech—an opinion that must be accepted as of the highest authority—is by no means so direct and unconditional as the above. It is as follows [2] : —

'The shallow surface-roots of the Beech mostly ramify throughout the loose upper layers of soil that are rich in humus. Under normal circumstances the suction-roots are not here provided with hairs, but are covered with a fungus (*Mycorhiza*), which acts in the capacity of a go-between with respect to water and food-supplies, and by means of which the trees derive much better nutriment from the layers of soil rich in humus, than is obtainable through the root-hairs from the mineral soil (B. Frank). A very valuable attribute of the Beech is that, judging from my numerous investigations as to the influence of woodland growth on the percolation of soil-moisture and on the chemical compo-

---

[1] *Op. cit.,* p. 266.
[2] *Forstlich-naturwissenschaftliche Zeitschrift,* 1893, p. 237.

sition of the air contained within the soil, in consequence of the ramification of its roots it keeps the soil more open or porous than the Spruce does. *To this property, combined with the circumstance that it is a shade-bearing species of tree, which yields a more nutritious and more easily decomposible humus than the conifers, it owes its soil-improving capacity and its great importance for mixed forests.'*

Where aeration is insufficient, the same practical effects take place as in the case of inundations. The plants consume the available supplies of oxygen within reach of their root-systems, and then sicken and die off ultimately, although of course different kinds of plants show great specific differences with regard to their accommodative power in this respect.

Nitrogen (N) is essential to the formation of the albuminoid substances which play so important a part in relation to the growth of plants. When organic matter decomposes, the nitrogen becomes sooner or later transformed into ammonia (NH_3), which again in turn becomes transformed with the aid of a fungus (as Winogradski has proved) into **nitric acid or hydrogen nitrate** (HNO_3). For the development of this fungus (*Bacillus*), or main intermediate agency, a certain amount of oxygen is requisite; but when there is a deficiency of oxygen in the air contained within the soil, **nitrous oxide or nitrogen monoxide** (N_2O) and free nitrogen are produced. This nitrification is confined to soil within about 18 inches of the surface. Other sources open to plants for obtaining the requisite supplies of nitrogen are the ammonia contained in the atmospheric precipitations, and also, but under special circumstances only, the free nitrogen of the air. Observations have shown that the summer showers contain a higher proportion of nitrogenous combinations than those that fall in autumn or winter. Although the older agricultural chemists, led by Boussingault, and strongly supported in 1861 by the Rothamstead experiments of Lawes and Gilbert, denied that plants could directly utilize the free nitrogen contained in the air[1], yet this was strenuously com-

---

[1] Compare Prince Kropotkin's article *On Recent Science* in the *Nineteenth Century* for August 1893, pp. 205 *et seq.*

bated by Ville all along, and since 1888 it has been abundantly proved by Hellriegel, B. Frank, Wilfarth, and others, that this can and does take place indirectly, more particularly in the case of *Leguminosae* (e.g. in the *Acacia* among our common trees). So long, however, as a supply of nitrogenous food is available in the soil, this capacity, for which the symbiotic aid of a *Bacillus* (*Bacillus Radicicola* or *Rhizobium leguminosarum*) is requisite, is not called into play [1]. The researches of these last-named scientists during the last decade have shown that all leguminous plants possess, by means of their root-nodules or tubercules, which contain these bacteria, the power of withdrawing from the air a sufficient supply of nitrogen for the requisite formation of essential albuminoid substances when these are not otherwise easily obtainable. According to Frank's researches also [2], the Alder is endowed with similar capacity to the *Leguminosae* for the formation of nitrogenous and albuminoid substances ; for the tubercules on its roots are found to contain an albuminoid protoplasma filled with the germs of a bacteroid fungus. Through a symbiotic process quantities of albumen are produced within these root-nodules, which become gradually absorbed into the tree and utilized for general requirements. Alders therefore thus appear, like the *Acacia*, able to derive supplies of nitrogen from the air, when requisite, through this symbiotic aid. So far, however, as woodland crops are concerned, though any abnormal increase in the ordinary supply of nitrogen obtained from the soil might probably exert great influence on the extent of seed-production and the frequency of mast-years, it certainly would not stimulate the increment in timber [3]. The precise manner in which woodland trees obtain the supplies of nitrogen requisite for their development has not yet been satisfactorily determined. But as woodland soils contain at best very small amounts of nitric acid, and are

[1] Ramann, *op. cit.*, p. 306.
[2] Report of the *Deutsche Botanische Gesellschaft* for 1891, p. 250.
[3] R. Hartig, *op cit.*, p. 221.

often totally deficient in it, their wants must be met by the nitrogenous compounds set free on the decomposition of organic matter within the soil ; and, even from this alone, one important function of the invaluable **humus** or leaf-mould can easily be seen. The warmer a soil, the more thorough its aeration, and the more its supply of moisture is removed from either of the extremes of dryness or wetness, so much the more rapidly and successfully will the nitric acid (so essential for agricultural crops especially) be evolved from decomposing organic remains. But as nitric acid is apt to be easily washed out of porous soils, this partially explains why loose sandy soil is less fertile than loams and other soils of a more retentive nature.

Water (H_2O) is also one of the essentials to the support of plant life. It not only forms a very direct source of nutriment, but it is also requisite for the solution of inorganic substances (*nutrient salts*), which can only be absorbed by the root-hairs when held in solution. By far the greatest portion of the water that is imbibed is transpired through the foliage. A certain quantity, however, is utilized in the formation of the organic tissue, the requisite hydrogen being exclusively obtained from this source ; whilst, during the process of assimilation, the organic substances are formed from water and carbonic acid by the elimination or setting free of oxygen (when $CO_2 + H_2O$ becomes $CH_2O + O_2$ &c. : see note 1 to page 71).

The requirements of woodland crops as to water vary greatly according to the species of tree, and for one and the same species according to the nature of the soil, the climate, and the exposure of any given locality. The finer the particles of soil, and the richer the admixture of humus, the more moisture can a soil contain. In situations with humid atmosphere the spiracles of the leaves open much wider than in dry air ; for the leaves have a certain power of accommodating themselves to the temporary conditions of the atmosphere so as to stimulate transpiration during a humid, and to retard it during a dry, state of the air. They are thus endowed with a certain capacity for regulating and maintaining as steady a rate of transpiration as possible. For all kinds of woodland crops, however, there is on the one hand a *minimum* supply of water requisite for thriving and for healthy development, and on the other a *maximum* normal limit beyond which any surplus of soil-moisture is prejudicial to their well-being. Although, with decreased transpiration, the process of assimilation may not be essentially interfered with, yet the imbibition of soluble salts is considerably decreased, and the amount of ashes or mineral constituents contained in the wood is less than in timber produced in a drier atmosphere.

Observations relative to the amount of water transpired by woodland crops were made by von Hönel in experiments made in 1879–1881. He found that, with the exception of the deciduous Larch, conifers transpire on the average only about one-sixth to one-tenth of the quantity of water that is evaporated by broad-leaved trees[1].

Taking the average of these three years, his experiments led to the result that the average transpiration of water per 100 grammes (0·22 lbs.) of dry leaf-substance amounted during the active period of vegetation annually to the following number of Kilogrammes (2·204 lbs.) :—

Broad-leaved Trees.		Conifers.	
	Kilogrammes.		*Kilogrammes.*
Aspen	95·970	Larch	120·234
Alder	93·300	Spruce	13·501
Lime	88·340	Scots Pine	9·426
Ash	85·615	Silver Fir	7·179
Birch	81·433	Black Pines	6·735
Beech	74·859		
Hornbeam	73·107	Mean average for all trees	69·800
Elm	66·170	Mean average for broad-leaved trees	82·520
Sycamore	58·596		
Oak	54·572		
Maple	53·063	Mean average for conifers	11·307

He also found that, whilst in conifers transpiration is greatest from foliage fully exposed to insolation, in broad-leaved trees it was greatest from foliage growing in shade, as the following table shows :—

[1] See Ramann, *op. cit.*, pp. 309–312.

	Kilogrammes per 100 grammes of dry leaves.	
	In the sun.	In the shade.
Beech	76·180	107·800
Hornbeam	81·300	98·900
Sycamore	61·690	76·190
Scots Pine	19·150	5·020
Silver Fir	13·910	4·850
Black Pines	8·760	5·250

As, however, the normal density of foliage, natural to the different kinds of woodland trees, varies greatly with each species, and also for one and the same species of tree according to the nature of the soil, the age and the method of treatment of the crop, and various other factors of sylvicultural importance, these data, which are by no means complete or intended to be advanced as finally authoritative, are rather of theoretical interest than of practical sylvicultural value. When investigating the actual quantity of water required by the Beech for normal transpiration annually, von Hönel estimated that—

A Beech tree 35 years of age
consumed per diem......... 1 kilo. = 2·204 lbs. = 0·220 gals. of water.

A Beech tree 50–60 years of
age consumed per diem ... 10 kilos. = 22·040 lbs. = 2·204 gals. of water.

A Beech tree 115 years of age
consumed per diem......... 50 kilos. = 110·200 lbs. = 11·020 gals. of water.

Computing the average number of trees per acre, he arrived at the following results :—

Age of pure crops of Beech.	No. of trees per acre.	Amount of Water consumed Annually.	
Years.		*Kilogrammes.*	*Gallons.*
35	1,600	280,000	61,712
50–60	520	920,000	202,768
115	160–240	1,400,000 to 2,160,000	308,560 to 476,064
Hence a summer rainfall of 12 inches suffices to meet these requirements.			

But it should be remembered that the quantity of soil-moisture requisite for transpiration through the foliage must be very considerabiy less in our damp insular climate than in the inland tracts of the continent of Europe with their much drier climate and consequently enhanced activity of transpiration.   There are many places along the east coast of England where even the total average annual rainfall is less than 12 inches, and where timber crops are still capable of thriving perfectly well.

Of much more practical interest than von Hönel's experiments must be considered the investigations that have more recently been made by Ebermayer with a view to determine the **relative transpiration** of the different kinds of forest trees, and the results of which correspond better than von Hönel's with the facts of actual sylvicultural experience.   Ebermayer analyzed foliage for the purpose of determining this by means of the percentage of water contained in the fresh leaves, and of pure mineral ash obtained afterwards from the combustion of the dry leaf-tissue.   The former table is now given, whilst that relating to the mineral ash will be referred to later on.   It need only be remarked that, as the nutrient salts are imbibed in a weak solution, and as the water is mostly transpired through the foliage in the form of aqueous vapour, whilst the salts eliminated are to a great extent deposited within the leaf-tissue, a steady and normal degree of transpiration is essential for the due conduct of the nutritive processes and the well-being of the trees ; but during sunlight, and also in warm dry situations, the rate of transpiration is of course greater than in shadow, or in cool, damp exposures, and at high localities where the relative humidity of the air is great.   Hence, *coeteris paribus*, it seems justifiable to conclude that leaves, which show large accumulations of pure mineral ash within them, not only make larger demands for mineral nutrients per unit of weight of foliage, but also at the same time transpire more water through their leaves in order to concentrate these deposits.   For the solutions taken up by the rootlets of various kinds of trees do not appear to vary much as regards strength (except in localities like Alder swamps where the solutions are abnormally weak), although of course in fertile soils (or on manured land) the solutions are richer than in land of inferior quality.   Proceeding from this point of view, Ebermayer made the following classification of trees with regard to their **relative power of transpiration** or **relative requirements as to soil-moisture**, by tabulating them according to the percentage of water the fresh leaves contained as compared with their total weight.

Broad-leaved Species of Trees.			
I.		**II.**	
Italian Poplar	70	Oak	63
Horse Chestnut	65	White Alder	63
Ash	65	Elm.	63
Common Alder.	64	Lime	63
Acacia (*Robinia*)	64		
III.		**IV.**	
Birch	61	Beech	57
Mountain Ash	61	Hornbeam	57
Sycamore	61	Silver Poplar	55
Willows (in general). . .	60		
Maple.	59		
Aspen.	59		

Coniferous Species of Trees.					
I.		**II.**		**III.**	
Cembran Pine .	64	Scots Pine . .	59	Silver Fir . . .	55
Black Pines . .	63	Spruce. . . .	59		
Larch	62	Mountain Pine .	57		
Weymouth Pine	61				

Taken on the whole, these scientific conclusions correspond fairly accurately with the knowledge acquired by sylvicultural experience, although at the same time certain kinds of trees, like Acacia, Birch, and Pines, are better fitted than other trees to accommodate themselves to conditions not exactly satisfying their normal requirements with regard to moisture.   It cannot fail to be noted from the above that long-needled Pines transpire more freely than the true Firs having short foliage. But in the conifers (which, as has been previously remarked, require only about $\frac{1}{8}$ to $\frac{1}{10}$ as much water as broad-leaved trees) the rate of transpiration is limited very considerably by their thick, strongly-cuticularized epidermis impregnated with resin and wax. Hence, not only the rate of transpiration, but also the movement of the sap upwards within conifers is much more sluggish than in deciduous trees like the Oak, Ash, Elm, &c. And whilst the needles, with their weaker transpiratory capacity, contain as much water as the leaves of many kinds of deciduous trees, the woody tissue of conifers contains more water than that of the latter.

The Indispensable Mineral Ingredients in the food-supplies of plants comprise Potash, Lime, Magnesia, Iron, Sulphur, and Phosphorus as well as Nitrogen. And at the same time Silica (often in large quantities), Chlorine, Soda, Manganese, and occasionally Alumina, are also to be found in the ashes of timber, after Carbon, Hydrogen, Oxygen, and Nitrogen have been eliminated by combustion; but the former can hardly be considered indispensable constituents. The different uses to which these mineral substances are put in the vegetable economy have not yet been quite satisfactorily determined. **Potash** (K_2O) is found chiefly in the foliage and the parts of more recent growth, which are concerned with the active processes of assimilation and of vital activity. **Lime** (CaO) is supposed to play a very important part in connexion with the formation and distribution of the carbo-hydrates, and in the formation of insoluble compounds with oxalic and similar injurious acids which are thus rendered harmless to the tender tissues of the plant. Whilst Beech and Black Pines show a decided preference for soils rich in lime, it has been proved by Fliche that even a slight percentage of carbonate of lime seems to act as a poison in the case of the Sweet Chestnut and the Maritime

Pine[1]. **Magnesia** (MgO) is utilized in the formation of albuminoid substances, and is supposed to have an important influence with regard to the production of seed. **Iron** (FeO and Fe_2O_3) is requisite, though only in small quantities, for the formation of the chlorophyll in the foliage, by means of which the work of assimilation is carried on with the aid of sunlight. **Sulphur**, taken up in the form of sulphates, i. e. salts containing sulphuric acid[2], is, like magnesia, employed in the formation of albuminoid substances, for which also **Phosphorus**, in the active form of phosphoric acid (P_2O_5), is most essentially requisite. The presence of **Chlorine** (Cl) and of **Soda** (Na) is most pronounced in plants growing near the sea-coast and on brackish soil. The specific action of these minerals is not very clearly understood; but they are supposed to assist in the movement of carbo-hydrates within the plant. For the practical purposes of Forestry the most important soil-nutrients are water, then lime, potash, and phosphoric acid; and the chief nitrogenous compounds are nitrates and ammoniacal salts or certain nitrogenous organic constituents in **humus**.

On soils which offer to woodland crops richer supplies of food than can be thoroughly assimilated by them, deposits of mineral substances take place within the plants in excess of their physiological requirements. Wherever this occurs in any excessive degree, it leads to a condition at once predisposing to disease and, at the same time, capable of offering least resistance to any attacks that may be made by fungoid parasites or by insect enemies.

---

[1] *Annales de la Station Agronomique de l'Est*, 1878, pp. 3 et seq.

[2] A common impurity to a slight extent in the atmosphere is sulphurous acid, which is carried by rainfall into the soil, and oxidizes into sulphuric acid ($SO_2 + H_2O$ becomes H_2SO_4). When present in the air in anything like excess, as in large towns and at all populous and manufacturing centres, where sulphurous acid is evolved on the combustion of mineral coal, it affects the thriving of trees. The sulphurous acid is imbibed by the foliage, and, on rain falling, becomes transformed into sulphuric acid; this rapidly acts as an irritant poison with regard to the internal, soft, cellular tissue of foliage.

The demands of the different species of woodland trees vary greatly with regard to mineral nutriment, as can be proved easily by an analysis of the amount of mineral matter found in the ash yielded on combustion and elimination of the non-mineral portions of their tissue (formed by Carbon, Hydrogen, Oxygen, and Nitrogen). The younger parts of the organism, and especially those actively engaged in the work of circulation, assimilation, and secretion (the **foliage** and the **cambium**), yield more ash because they are richer in salts of potash, phosphates, and nitrogenous compounds, than the older portions (the **alburnum**), and the lifeless tissue in the interior of the stem forming the hard heartwood (the **duramen**).

The leaves are much richer in mineral matter than any other part of the plant; whilst the bark contains more than the wood, and the younger portions of the tree contain more than the older portions. When parts of the tree are about to die off (as, for example, in the case of the foliage of deciduous trees in autumn), the more important classes of mineral nutrients—Potash, Phosphoric Acid, Magnesia, Lime — are transferred into the still active portions of the organism, to which they wander back together with albuminoid substances and soluble carbo-hydrates.

The total proportion of mineral matter in the timber of broad-leaved genera of trees is seldom under $\frac{1}{3}\%$ or exceeding $\frac{1}{2}\%$ of the dry substance of the wood. It usually ranges between o·3 and o·45%; but, in the case of Birch and of conifers, it varies from about $\frac{1}{8}$ to $\frac{1}{3}\%$ (o·17 to o·27). The heartwood of old stems contains less ash than the sapwood. It is also comparatively devoid of albuminoid substances, and is therefore more durable, i.e. it is less exposed to the disintegrating and decomposing influences of alternating damp and heat, or to the organic attacks of saprophytic fungoid growth and insects.

According to the demands of the different kinds of trees for food-supplies, their wood shows the following percentages of mineral matter per 100 parts of dry tissue, according to Ebermayer[1] :—

Class of Trees.	Wood of the	Percentage of Ash.	Potash.	Lime.	Phosphoric Acid.
Deciduous	Stem .	0·33–0·94	0·06–0·16	0·10–0·65	0·03–0·06
	Twigs and small branches	0·75–2·00	0·16–0·46	0·21–1·39	0·12–0·30
Coniferous	Stem .	0·31–0·45	0·04–0·15	0·06–0·14	0·011–0·027
	Twigs and small branches	1·01–2·30	0·11–0·40	0·25–0·44	0·10–0·22

In the majority of plants cultivated by the farmer the quantity of pure ash left after the combustion of the dry tissue varies from 6 to 8% of the total weight.

And, according to the same high authority, the annual demands made by pure crops of forest trees, for the formation of wood only, are per hectare in kilogrammes, or *approximately* per acre in lbs. avoirdupois, as under :—

Species of Tree.	Age of Crop.	Total Quantity of Ash.	Potash.	Lime.	Phosphoric Acid.
		years	lbs.	lbs.	lbs.	lbs.
Scots Pine	80–100	12–16	2–3	7–11½	⅞–2
Spruce . .	100–120	23–24	4–5½	9–11	1½–2½
Silver Fir .	90	39	10	4¾	3
Beech . .	90, 100, & 140	30–49	6½–8¼	15–20	2½–4¼
Hornbeam	40	30	3½	20	2¼
Common Alder . .	70	18	2	12	1½
Birch . .	50	12½	2⅛	4	1⅛

And in investigations in which all previous thinnings were left out of calculation.

Beech . .	50	23½	5	11¼	1⅛
Oak (*Q. sessiliflora*) .	.	27⅛	3	20⅛	1

[1] *Forstlich-naturwissenschaftliche Zeitschrift*, 1893, pp. 225, 226.

In trees on which the bark long remains thick and fleshy, larger quantities of mineral deposits are stored up in the rind than in those which tend to the early formation of a rough and rugged cortex.

As the great bulk of the mineral matter is to be found in the foliage of trees, it therefore follows that woodland crops must make the greatest demands on mineral nutrients just at the time when they develop the greatest density of leaf-canopy— that is to say, about the twentieth to thirtieth year for Scots Pine, the thirtieth to fortieth year for Spruce, and the fortieth to fiftieth year for Oak and Beech, on the better classes of soil, or perhaps about a decade later on soils of poorer quality. On the inferior classes of land it is therefore just about these ages that crops of those particular kinds of trees will be most likely to show signs of the soil not being able to supply easily the normal demands made for fully adequate supplies of nutrient salts.

But, at the same time, the absolute demands in this respect vary enormously in the different genera and species of trees. For the production of any given volume of wood, Weymouth Pine, Scots Pine, and Birch make relatively the smallest demands on mineral nutriment; after these follow Alder, Spruce, Silver Fir, Hornbeam, Beech, and Oak; whilst Ash and Acacia make the greatest demands.

The quantities of the different kinds of mineral nutrients, required annually by the various species of woodland trees for their normal production of wood, range between wide limits for the different individual kinds of trees, as the above table shows. But mere analyses of the timber alone will give no useful key to the total annual requirements; for, as has already been stated, the most of the mineral salts absorbed from the soil are finally stored up or deposited in the foliage during the process of assimilation, and are returned again to the soil after the fall of the leaf. It may, however, in passing, be again briefly remarked that, for the production of wood only, the demands vary as follows in high-forest crops of various sorts :—

With regard to Lime   from 4 to 20 lbs. per acre per annum.
      ,,   Potash   ,,   2 to 10   ,,   ,,   ,,
      ,,   Phosphoric Acid ,, ⅜ to 4½   ,,   ,,   ,,

In order to arrive at more definite estimates than these investigations alone could yield, Ebermayer made analyses for the purpose of determining the amount of mineral ash contained in the dried foliage of the more important kinds of forest trees, and obtained the following results as to the percentage of pure ash deposited in the dry leaf-tissue.

A. DECIDUOUS, BROAD-LEAVED TREES.

I.	Exacting.	Ash	7·61
		Italian Poplar . . .	7·04
II.	Making considerable demands.	Elm	6·92
		Lime	6·55
		Mountain Ash . . .	6·20
		Acacia (*Robinia*) . .	6·16
		Horse Chestnut . .	6·03
III.	Making moderate demands.	Sycamore	5·33
		Maple	5·21
		Aspen	5·15
		Willow	5·11
		Oak	5·10
IV.	Making slight demands.	Silver Poplar . . .	4·69
		Hornbeam	4·20
		Beech	4·02
V.	Least exacting.	White Alder . . .	3·92
		Birch	3·90
		Common Alder . . .	3·69

B. CONIFEROUS TREES.

I.	Making greatest demands.	Silver Fir	2·93
		Cembran Pine . . .	2·78
		Larch	2·54
II.	Making moderate demands.	Spruce	2·39
		Weymouth Pine . .	2·35
		Scots Pine	2·28
III.	Making least demands.	Black Pines. . . .	1· 2
		Mountain Pine . . .	1·36

From these tables it appears that none of the conifers make anything like such large demands as the broad-leaved genera of trees **per unit of foliage**; for the most exacting of the former makes considerably less demand than the least exacting of the latter. But the amount of nutrients annually withdrawn depends on the total quantity of foliage and of timber formed annually; and, as yet, no exact data of this description have been collected and tabulated. As a matter of actual experience, however, we know that soil improves greatly under coniferous crops, so long as a good canopy of foliage is maintained; hence there is good reason for believing that, in making very moderate annual demands, they enable the capital in nutrients to be increased within the soil year by year. But the very favourable physical influences exerted simultaneously on the soil by means of the canopy of evergreen foliage overhead must not be left out of the reckoning when this matter is being considered.

It will be noted, by a comparison of the above table with that previously given for requirements as to water, that, in a general way, the trees transpiring most freely, also withdraw the largest mineral food-supplies from the soil. And these results correspond very fairly with actual sylvicultural experience. Where discrepancies seem apparent, a study of all the circumstances connected with the growth of the tree

in question will soon explain the paradox. Take the case
of the Acacia (*Robinia*) for example.   The analysis of its
foliage distinctly shows it to belong to the **exacting** class of
trees; and yet it is to be seen in vigorous growth on soils
of inferior quality.   It must, however, be borne in mind that
the total leaf-production of the Acacia is not heavy; hence,
even on poor classes of soil, it has less difficulty in supplying
its higher requirements **per unit of weight of dry leaf-substance** than other genera having more moderate requirements
in this latter respect but endowed with a heavier total crop
of foliage.   And that, in the case of this particular tree, the
symbiotic aid of the *Bacteria*—which are to be found in the
nodules of all *Leguminosae*, and which are capable of assimilating nitrogen from the free nitrogen of the air—may possibly
be of assistance in supplying nitrogenous food, to be used
for the easier assimilation of other mineral food, is not at all
difficult of comprehension.

In the case of the Ash, again, which makes the greatest
demands of all the woodland trees as to mineral food-supplies
per unit of weight of dry leaf-substance, this really withdraws
only a moderate actual quantity of nutrient salts from the soil,
as its total amount of dry leafy substance is slight compared
with that of many other genera of trees, which require smaller
quantities of nutrients per unit of dry leaf-tissue, but yet withdraw far larger total quantities from the soil annually. Subject,
however, to the **Law of the Minimum**, referred to at the opening of this chapter, the productive capacity or fertility of any
soil is dependent on the quantity of nutrients available within
it in a form that can be absorbed by the rootlets, and on the
extent to which their absorption is favoured by the combination
of the various physical factors determining its tenacity, moisture,
warmth, and depth.   But every soil is not, as one of the earlier
sylvicultural chemists, Gustav Heyer, asserted [1] adapted to the

---

[1] *Forstliche Bodenkunde und Klimatologie*, 1856, p. 3.   He assumed
that almost any soil could produce any given kind of timber, if it only had
the requisite amount of moisture.

growth of every kind of forest tree ; for the supply of nutrients available in a soluble form varies with the chemical and the physical composition of the different classes of rocks, the degree to which disintegration and decomposition have advanced, the depth of the soil permeable by the root-systems, the amount of organic matter contained in the soil, and the physical properties with which this is endowed. And the want of any *one* food-material, or perhaps even an unfavourable condition of *one* physical factor (more especially one of those relating to moisture or depth), may render a soil unsuitable for the normal development of any given kind of tree. But, undoubtedly, the inferiority of many woodland soils may also be due to poverty in one or other of the essential soluble nutrients, though this is generally combined at the same time with unfavourable physical conditions. The object of scientific sylviculture is, therefore, to plant only such trees as are most likely to attain normal development, and to yield the best monetary returns, without exhausting the soil or allowing its physical properties to become prejudiced or impaired by want of protection from sun and wind.

From this it follows, theoretically, that the cultivation of **coppice-woods**, in which the crop is formed entirely of small branches containing relatively larger quantities of mineral substances than are to be found in mature wood, must tend to exhaust the easily available supplies of lime, potash, and phosphoric acid sooner than crops grown under **high-forest**. And this theoretical deduction coincides exactly with the results of practical experience; for it is well known that Oak-coppice for tanning-bark, or Osier-holts for withes, should only be made on good land, and that if inferior soil be cleared of Oak-coppice every twelve to sixteen years, it runs great risk of deteriorating. The treatment of timber-crops in **high-forest** is, as long sylvicultural experience has clearly shown, the most conservative manner of utilizing the soil; whilst **copse** or **coppice under standards** stands about midway between the two other systems, as being less exhausting than coppice, though more exacting than high-forest.

High-forests of timber can never lead to exhaustion or deterioration of the soil, so long as this remains protected by a sufficient leaf-canopy, and is not robbed by wind, or otherwise, of the natural manure obtainable through the decomposition of the dead foliage. On the contrary, and more especially in the case of trees like Scots Pine and Spruce, which make lower demands for potash and lime than other species, it is almost always the case that the nutrient salts contained in an available form within the soil at the end of the period of rotation, i. e. on the fall of the mature timber, are considerably in excess of the quantity that was present at the time of such crop being formed—provided always that the fertility or productive capacity of the soil has been duly safeguarded by the maintenance of a good, unbroken canopy of foliage throughout all the life-periods of the crop. For this reason, woods formed of trees like Pines, Spruces, and Firs, recommend themselves for the recuperation of the productive energy of soils that have been allowed to deteriorate by being kept long under lightly-foliaged kinds of trees, like Oak, Ash, Elm, or Larch; because the broken canopy formed by these light-demanding species is unable, without the aid of undergrowth, to protect the surface-soil from the exhausting influences of sun and wind, and from having the accumulated supplies of humus consumed unprofitably by a more or less . rank and unremunerative growth of weeds.

CHAPTER V

SOIL AND SITUATION IN RELATION TO WOODLAND GROWTH

I. Soil in Relation to Woodland Growth.

By **Soil** is understood the product of the decomposition of
the rocks forming the crust of the earth. Soil is formed
mechanically and physically by rains, frosts, rivers, volcanoes,
&c., or chemically and organically by oxidation, deozidation,
and changes effected by carbonic acid. Through its permeability and its other physical factors, the soil enables trees
to develop the root-systems necessary alike for their support
mechanically and physiologically. By far the greatest portion
of the food of all plants possessing chlorophyll is obtained
from the carbonic acid of the atmosphere; but certain necessary substances (nutrient salts) are only obtainable from the
soil when held there in solution, so that they may be imbibed
by the rootlets. A distinction is made between the **soil** or
surface-soil and the **subsoil.** The former consists of the
usually more thoroughly decomposed earth forming the upper
layer permeated by the root-system; whilst the latter may, or
may not, be of similar composition to the superposed layer.
Even when such patent signs of shallowness of soil as stony
outcrops are not visible, the want of depth may easily be
noted by the stunted appearance of the timber-crops, in which,
owing to the deficiency in soil-moisture and food-supplies, an

abnormal development of the root-system takes place relatively
to the ascending axis, and to the crown of foliage charged
with the duties of transpiration and assimilation. The chief
signs of depth of soil are length of bole, fuller coronal development, and richer soil-covering of underwood, brushwood, and
weeds. A tolerably correct estimate may, in fact, be obtained
concerning the depth and the general quality of the soil from
the mean average height of the timber crops on such soils as
are usually to be found under woods. Thus Grebe gives the
following averages for the Thüringer Wald in the middle Saxon
states [1] :—

AGE OF FOREST IN YEARS, AND AVERAGE HEIGHT IN FEET.

Kind of Forest.	Quality of Soil.	Yrs. 40	Yrs. 50.	Yrs. 60	Yrs. 70	Yrs. 80	Yrs. 90	Yrs. 100	Yrs. 110	Yrs. 120
		ft.	ft.	ft.	ft.	ft.	ft.	ft.	ft.	ft.
Spruce . .	I.	43	55	69	81	90	97	103	107	109
	II.	37	48	60	69	77	84	88	91	93
	III.	30	40	49	58	64	69	72	75	76
	IV.	23	30	39	45	50	54	57	58	—
	V.	16	21	27	32	36	39	40	—	—
Scots Pine	I.	39	52	65	75	84	90	95	98	96
	II.	36	47	58	67	73	79	83	85	98
	III.	31	41	49	57	62	66	68	70	—
	IV	25	33	40	45	48	51	53	54	—
	V.	19	25	29	33	36	37	38	—	—
Beech . .	I.	34	48	59	69	78	84	89	92	94
	II.	29	41	50	59	67	73	76	79	81
	III.	24	33	42	49	57	61	64	66	67
	IV.	19	27	33	40	44	48	50	51	52
	V.	(Not grown on such inferior soil.)								

How the general quality of the soil affects the outturn in
timber from the different kinds of woodland crops may be seen

[1] *Die Betriebs- und Ertragsregelung der Forste*, 2nd edit. 1879. These
heights are of course only rough approximations so far as growth of these
trees in Britain is concerned. But there is no climatic reason why they
should not be equalled, or even excelled, in properly managed woods.

from a reference to the tables on pp. 43-45 of the author's *British Forest Trees* (1893).

The **Chemical Composition** of any soil is in so far of importance to tree-growth that each of the necessary mineral constituents, always found forming part of the ash to which timber may be reduced by combustion, must be present in it in sufficient quantity in a soluble form. The adaptability of any given soil for the growth of any particular species of woodland crop is determinable by the **Law of the Minimum**, according to which the minimum amount of any one essential constituent of the necessary food-supply in the soil limits its total productive capacity for a given kind of timber. Thus, for example, Oak and Beech extract very much more potash and phosphoric acid from the soil for the production of their timber than Scots Pine or Spruce ; hence the latter may thrive well on soils where (according to the law of the minimum) the former are of exceedingly indifferent growth owing to the comparative scarcity of these ingredients.

As a matter of fact most soils contain the essential constituents of tree-food in sufficient quantities to maintain tree-growth of any kind. But as they are not always held in solution, they are not continuously available for imbibation by the root-system ; for it is only when occurring in the form of **soluble salts**, that the nutrients can be absorbed by the suction-rootlets. It was from this point of view that Gustav Heyer asserted that almost any soil could produce any given kind of timber, provided that it contained the requisite degree of moisture—an assertion which, as has previously been shown, is only approximately, but not scientifically correct.

Sylviculture varies essentially from Agriculture in being less directly dependent on the fertility or inorganic richness of the soil. Though it is quite certain that rich soils, yielding too copious supplies of food-material to be assimilated thoroughly, would produce large quantities of timber, yet it would be of a soft spongy nature, little able to withstand either the attacks of fungoid diseases whilst growing, or of red and white rot after conversion. According to Weber[1], the analyses of different woods show that the composition of timber consists, on a rough average, of 50 % Carbon, 42 % Oxygen, 6 % Hydrogen, 1 % Nitrogen, and 1 % Ash only, for a great many of the mineral constituents abstracted from the soil are deposited in the foliage during the process of preparing the food-supplies for assimilation, and are returned again to the soil when defoliation occurs, and the dead foliage decomposes into **humus** or mould. But the whole demand for mineral food made by woodland growth is, roughly speaking, only about one-half of what it amounts to in the case of agricultural crops, according to the investigations made by Ebermayer[2], which may thus be tabulated :—

Average of	Total quantity of ash.	Silica. SiO_2.	Potash. K_2O.	Lime. CaO.	Magnesia. MgO.	Phosphoric Anhydride. P_2O_5.	Sulphuric Anhydride. SO_3.	Other constituents.
Mixed agricultural crops	235	37	78	43	17	28	11	21
Woodland growth —timber & leaves	126	29	11	62	10	8	3	3
Woodland growth —timber only . .	19	1·6	4	9	2	1·4	0·4	0·6
Approximate ratio of sylvicultural demands to agricultural needs . .	$\frac{1}{2}$	$\frac{3}{4}$	$\frac{1}{7}$	$\frac{3}{5}$	$\frac{3}{5}$	$\frac{2}{7}$	$\frac{1}{4}$	$\frac{1}{7}$

As, in sylviculture, the greatest demands are for lime and silica, constituents abundant in every soil, and, as by the fall of the leaf, most of the mineral food is returned to the soil again, it will be at once apparent that woodland growth does not

[1] *Die Aufgaben der Forstwirthschaft* (Lorey's *Handbuch*), p. 62.
[2] *Physiologische Chemie der Pflanzen*, 1882, vol. i. p. 761.

exhaust the land to anything like the same extent as agricultural crops, which require manuring in order to supply deficiencies, and to stimulate the upper soil to increased decomposition. And when the dead foliage is retained on the ground and decomposes to form humus, this in turn exerts a very beneficial hygroscopic influence, stimulates the soil to more rapid decomposition, renders the mineral food-materials more easily soluble, and, in short, improves the productive capacity of the land, instead of exhausting it. This shows the importance of maintaining a close leaf-canopy, and of adopting all practical measures in order to obviate insolation of the soil, and prevent the free play of winds that may remove the foliage and exhaust the surface-moisture.

But the mineral constitution of the soil is not the only factor determining its productive capacity sylviculturally; for various physical factors are of equal, and often greater, importance— e.g. depth, porosity (aeration), consistency, &c. So far, however, we may be guided by practical experience as to assert that Elm, Maple, Sycamore, and Ash can only be grown satisfactorily on fertile soils; whilst Birch and Pines are least exacting as to mineral richness. Between these extremes, Oak, Beech, Lime, and Silver Fir make somewhat greater demands than Sweet Chestnut, Larch, Hornbeam, Spruce, Alder, Aspen, and Willow. When planted out on soils not naturally favourable to their growth and development, Scots Pine and Birch show a decidedly greater capability of accommodating themselves to any given conditions than Oak, Beech, Spruce, and Silver Fir; whilst Ash, Oak, Elm, Maple, and Sycamore—those genera whose demands on fertility are greatest—possess the least power of accommodation.

Soils are largely made up of **Salts**, with Oxides and sometimes Chlorides. The **Salts** are mostly Silicates (often combined with water), Carbonates, and Sulphates; whilst the Phosphates, so important for plant-life, only occur occasionally. In the formation of these **Salts**, the most important of the **Acids** are Silica ($Si O_2$), Carbonic acid (CO_2), Sulphuric

Anhydride or Sulphur Trioxide (SO_3) and Phosphoric Anhydride or Phosphorus Pentoxide (P_2O_5); whilst the chief **Bases** are Potash (K_2O), Soda (Na_2O), Lime or Calcium Oxide (CaO), Magnesia or Magnesium Oxide (MgO), Ferrous Oxide (FeO), Sesquioxide of Iron or Ferric Oxide (Fe_2O_3), Alumina (Al_2O_3), and Manganese Dioxide (MnO_2). The **Silicates** comprise quartz, serpentine, talc, felspars, mica, hornblende, augite, and chlorite; the **Carbonates** include carbonate of lime, chalk, and dolomite; to the **Sulphates** belong anhydrite and gypsum or calcium sulphate. When the water of crystallization in many salts composing rocks becomes heated and driven off, the salt crumbles into powder; and, at the same time, all rocks, being porous, are pervious to water, which not only dissolves certain salts, and holds them in solution ready for absorption by the roots of plants, but also effects transformations in many chemical compounds. Thus water impregnated with carbonic acid dissolves silicates containing alkaline earths and ferrous oxide, as also carbonate of lime ($CaCO_3$) and ferrous carbonate ($FeCO_3$) which are easily soluble in it, whilst magnesium carbonate ($MgCO_3$) is much less so. The products of the decomposition of rocks are mostly alkalies in combination with silica and carbonic acid, together with carbonate of lime, magnesia, and ferrous oxide, whilst soluble sulphate of lime is also found in most soils; but as these salts act and re-act on each other, many various compounds are formed.

Rocks of complicated structure decompose most rapidly, as, for instance, those rich in felspars, or in compounds of iron, in contrast with silicious slate or quartz-rock; pure limestone decomposes slowly, but this difficulty diminishes with any increase of alumina and iron. Sandstones and conglomerates decompose in proportion to the cementing matter they contain, and to the degree in which this is affected by moisture. Fine-grained and massive rocks are less apt to fissure and crumble than coarse-grained stones or slaty beds, as they afford fewer points of attack to water, mosses, herbage, and other disintegrating influences.

Vegetation plays no unimportant part in the formation of earth, not only by helping to fissure and cleave the soil and subsoil, so as to enable water to effect an entrance, but also by reason of the decomposition of the organic débris of foliage, dead wood, &c., and the formation of **humus** or mould, which takes place under the combined action of oxygen, moisture, and a minimum warmth of 52° Fahr. When the humification takes place only under partial exposure to the atmosphere, humic and similar acids are formed, and saprophytic fungi aid in the work of decomposition, by attacking the albuminoid substances first of all. These acids (humic, ulmic, geic, &c.) have a strong affinity for ammonia (NH_3), which is itself essential to the nourishment of trees, and which cannot be assimilated by the foliage from the free nitrogen of the air, although, as Hellriegel, Frank and others have shown, it can be obtained by the *Leguminosae* from the air circulating within the soil by means of symbiosis with a fungus (*Bacillus radicicola*) found in the nodules on the roots (see page 74).

Humus condenses gases and atmospheric moisture, and possesses a certain amount of warmth, partly owing to retention of atmospheric warmth received from sunshine, and partly to the generation of heat by the chemical process of decomposition. Thus, whilst mechanically making clayey soils less stiff, and sandy soils more binding, it also warms the former, and makes the latter less liable to be affected by the changes in the atmospheric temperature.

II. The Classification of Soils.

Any classification of soils according to their geognostic or geological origin would be very misleading, for (1) the same kind of rock does not always yield similar soil, (2) the qualities of the soil depend on the extent to which decomposition has taken place, and (3) some of the lighter products of decomposition (clay) are more easily washed away than others that are heavier (sand). Hence, for practical sylvicultural purposes, no better classification of soils can be adopted than that made by Grebe, which is now generally adopted at the experimental stations on the continent :—

Sandy Soils, containing 75 % or more of disintegrated sand. They occur as sand-drifts, or are the product of the decomposition of sandstones.

>*Sand* consists of about 90 % of sand, and not more than 10 % of clay and other constituents.

>*Loamy sand* consists of about 75 to 85 % of sand, and from 15 to 25 % of clay and other constituents.

Loamy Soils, always tinged with iron, containing 60 to 70 % of fine sand and silicious dust, the rest being chiefly made up of clay with less than 5 % of lime, and an almost constant quantity (5 %) of hydrated ferric oxide.

>*Loam* consists of about 60 % of sand and 40 % of clay.

>*Sandy loam* consists of about 70 % of sand and 30 % of clay.

Clayey Soils, containing 50 % or more of clay. They are mostly formed from rocks rich in feldspars, augite, and hornblende, of sandstones and conglomerates cemented with clayey cohesive substance, and of the clayey layers and bands throughout sandstones and lime formations.

>*Clay* consists of about 60 to 70 % of clay, the rest being chiefly sand, and also other constituents.

>*Loamy clay* consists of about 50 % of clay, and 50 % of sand.

Limy Soils, containing not less than 10 % of carbonate of lime. They are produced chiefly by limestones, but also by other rocks containing lime felspar (labradorite). It often happens that limestone rocks produce a loamy rather than a limy soil on decomposition.

>*Lime* consists of at least 50 % of carbonate of lime, the rest being chiefly clay.

>*Clayey lime* consists of about 40 % of lime and 60 % of clay.

>*Loamy lime* consists of about 30 % of lime and 70 % of loam. When rich in carbonate of magnesia it is called *dolomitic lime* or *dolomite.*

>*Marl* consists of 10 to 20 % of carbonate of lime, with 80 to 90 % of clayey and sandy ingredients.

Sandy soils have a coarse gritty feeling when slightly moistened. When there is any considerable admixture of loam, marl, iron, or humus, sandy soils are said to be loamy, marly, ferrugineous, or humose. Sands become easily heated or cooled, and have a poor capacity for retaining moisture and nutrient salts in solution ; hence they form the inferior qualities of soil, on which trees require a considerable growing-space in order to obtain the requisite supply of food, fail to maintain good leaf-canopy, become deficient in increment, produce little seed, and are otherwise deficient in reproductive capacity. Owing to their becoming easily warmed by day, they stimulate to the early germination of seed, and movement of sap, and flushing of foliage in spring ; but, as they cool rapidly at night, this exposes the young vegetation to danger from late frosts. Earthy ingredients, and humus in particular, improve the quality of sandy soils. When forests approaching maturity show good development on such soils, their natural regeneration is safe and simple ; but where the tree-growth is indifferent, it is advisable to resort to planting, as the young growth of seedlings is often too intolerant of shade to make natural regeneration under parent standards a success. In the planting up of such soils for the first time, Scots Pine is the best crop, mixed with whatever other species of trees may seem likely to thrive at all well, as, for example, Douglas Fir, Black Pines, Spruce, Menzies Spruce, &c.

Clayey soils stick if brought in contact with the tongue, and smell of ammonia if breathed upon; rubbed between thumb and forefinger they feel fatty to the touch, and take a polish or burnish if rubbed with the thumb-nail. They may be grey, yellow, or brownish-red in colour, and are usually tinged with iron. Their leading characteristics are tenacity, weak hygroscopic power, and impermeability to moisture ; but they are strongly retentive of moisture when once saturated. Such soils are cold, and are apt to become water-logged. Any considerable admixture of sand, iron, lime, or marl—forming sandy, ferrugineous, limy or marly clays—tends to modify these characteristics. As the soluble nutrient salts are not apt to be washed out of the soil, clays offer comparatively large supplies of tree-food within a small space; hence the crops have a greater density of leaf-canopy at all stages of their development than similar woods growing on sandy soils. Unless the subsoil be permeable, however, there is a danger of such tracts becoming marshy. Owing to their low conductivity of heat, woodland growth on clayey soils seems backward in spring. In the natural regeneration of mature crops a good deal of soil-preparation is usually required on clays in order to assist in the decomposition of the dead foliage, which often forms a thick layer on the soil in consequence of the density of the canopy overhead. Not only is more foliage thrown down, but this also retards the formation of humus by keeping the soil at a low temperature. On such soils natural regeneration is very easy; for the preliminary clearances stimulate to increased production of seed, and the young seedling growth bears shade better than on poorer sandy soils. The reproductive capacity of copse and coppice is also greater on clays than on sands.

Limy soils effervesce when tested with drops of nitric acid. Marls and true limes are both comprised within this group; but in the former the limy constituent is equally admixed throughout each particle of the soil, whilst in the latter it is not. These also become by admixture sandy, loamy, clayey, or stony limes or marls. Limy soils are apt to be wanting in depth; but with careful treatment they often show fine growth of timber. Their consistency favours the equable distribution of soil-moisture, even on steep hill-sides; hence the root-system easily attains normal development, and has few difficulties to overcome in satisfying its requirements in the way of food-supplies. But, when such soils have deteriorated through insolation following on bad management, the soil-moisture is not retained; the finer particles of earth get washed away rapidly; the soil rapidly becomes shallow, dry, and hot; and timber production diminishes considerably. The re-wooding of dry, limy hill-sides is one of the most difficult of sylvicultural tasks.

Some very good results have occasionally been attained by sowing out lucerne (*Medicago sativa*) and beech-nuts at the time of planting out Austrian Pine (*Pinus Laricio* var. *austriaca* Endl.), the lucerne being left to form humus in place of being utilized as fodder. A soil-covering of dead foliage and humus on limy tracts is essential for the maintenance of a productive capacity anything like commensurate with their mineral strength; hence natural regeneration is advisable when removing the mature crop. As the dead foliage rapidly decomposes, and seed-production is usually good on limy soils, little or no preparatory clearance is necessary previous to the seed-felling; whilst, as the seed is generally of good quality and high germinative power, and the seedlings are not apt to wilt or droop under the shade of the parent standards, the regenerative process is comparatively easy. But, when limy soils are sandy or stony, greater care must be taken not only in making the regenerative fellings, but also in all the operations of tending throughout the whole life-period of the crop. Limy soils that are properly protected against insolation are in every way well endowed with reproductive capacity as coppice.

Loamy soils have greater resemblance to clay than to sand, but neither feel fatty when rubbed between finger and thumb, nor take any definite polish when rubbed with the finger-nail. A general mildness is their leading characteristic; but they approach more to the one or other of the three main groups according to the amount of sandy, clayey, or limy admixture contained in them. Though their capacity for absorbing and retaining soil-moisture is determined to a great extent by the . nature of the subsoil upon which they rest, yet they are in general favourable to sylvicultural crops; and the species of trees selected for growth are usually chosen for financial reasons, or for reasons connected rather with the physical properties than with the mineral composition of the soil. At high elevations with humid atmosphere the Spruce or the Douglas Fir is recommendable as the ruling species on loamy soils; whilst at lower

levels the Oak is one of the species thriving best in mixed
forests of broad-leaved trees.   Natural regeneration is easy on
loamy soils when a good seed year comes round ; but, as good
seeding or mast-years can not be depended on in crops on
loam, it is best to sow seed procured from elsewhere rather
than to risk deterioration of the soil from insolation after
once the preliminary fellings have been made.

As the fertility or productive capacity of soils is neither
dependent on, nor mainly proportional to, the mineral richness
of the rocks from whose decomposition they are formed, no
hard and fast classification as to their mineral strength is
possible.   It can only be said that, in general, clayey soils are
slow in forming humus and giving nutrients to tree-life, whereas
limy soils decompose organic matter rapidly and stimulate
tree-growth, and that intermediate positions between these two
groups are occupied by sandy and loamy soils, except when
the sands are deficient in soil-moisture.

All soils are improved by an admixture of **humus** to a moderate extent.   It not only adds depth to the land, but, being
of a strongly hygroscopic nature, also condenses and retains
atmospheric moisture ; whilst, owing to its low conductivity, it
protects the soil against evaporation.   It is also directly active
in aiding the continuous decomposition of the soil by means
of the carbonic acid set free during the process of decomposition.   It modifies all extremes of physical properties in soils ;
and though not an absolute necessity for the production of
woodland crops, yet it is of inestimable value in stimulating
the action of soils, no matter of what geognostic origin.
Indeed, as Gayer remarks[1], there can be little doubt that
*'humus forms the most important factor relative to tree-growth, and
is a priceless treasure as regards the production of woodland crops.'*
And the beneficial influence it exerts on the aeration of soils
is by no means its least important quality.   When soils contain
20 % or more of humus they are termed **humose** ; fertile silt left

[1] *Waldbau*, 1889, p. 27.

after inundations is often of this character. Sometimes, as in the case of marshes, moors, and peat bogs, the accumulations of humose soil is so great that the roots of trees are unable to reach the mineral soil below it; such formations usually take place on sandy soil resting upon clay.

Ferruginous soils are coloured brownish-red, and contain at least 10 to 15% of ferric oxide or hydrated ferric oxide. But even by 2 to 5% of ferric oxide the general character of sandy soils may become affected, although considerably larger quantities may fail to effect any noticeable alterations in clayey or loamy soils. Thus moorpan, sometimes containing only about 2% of ferric oxide, is just as impenetrable to the roots of trees as iron-band or limonite with its 60 to 70%. In both cases tree growth is hindered until the impenetrable layer has been broken through, so as to allow of the roots reaching the subsoil, and of the moisture circulating normally.

III.   The Physical Properties of Soils.

To these, in a far greater degree than to the mineral composition of any soil, is due its capacity to sustain timber crops, and to yield to them in due proportion the various forms of nutriment requisite for their growth, development, and reproduction. All the physical properties act and react on each other in determining the quality of any given land.

Cohesiveness or Tenacity is the resistance offered by a soil to instruments used in separating its particles, or to force employed in disintegrating them. This property is not alone of importance in relation to air, moisture, and warmth, but also indicates the resistance to be overcome by the root-systems in penetrating into, and ramifying throughout, the soil.

Clay soils have the greatest cohesiveness, sand the least; lime approaches more to the clay, while loam resembles rather the sand. An admixture of humus tends to level all classes, making clays and limes less tenacious, and loams and sands more binding. The tendency of soils to expand with moisture, and to shrink on parting with it again, is practically proportional to its tenacity. Soils may be classified in this respect as—

Heavy, stiff, or tenacious soils, which fissure and crack deeply when dried up. Clays and clayey loams, limes and marls belong to this class.

Mild soils, exhibiting superficial cracks when suddenly dried up, but which when nearly dry have the power of retaining the form of clods. They are usually composed of favourable admixtures of clay, sand, and lime, as in loams, sandy loams, and loamy limes.

Light soils, capable of forming clods when moist, but when dry showing a strong tendency to disintegrate, as in the case of loamy sands and sandy marls.

Loose soils, which even when moist have little tendency to form clods. To this class belong the poorer sandy soils.

Shifting soil, liable in a dry condition to be carried away by the wind, as in the case of sand-drifts and dunes.

In a light soil, which is not deficient in moisture, humus, and mineral nutrients, a maximum of rootlets can be developed, and through these the maximum of timber. Hence one finds the greatest increase in cubic contents, and in particular the greatest average annual increment in height, on alluvial deposits having finely pulverized particles. But when loose soils are wanting in soil-moisture, planting operations are apt to be unsuccessful; for the plants have great difficulty in establishing themselves so as to send down their roots to lower levels and thus obtain an assured supply of the water requisite for transpiration through the foliage. Thus, on the sandy soils in some parts of Sussex, for example, seedling Oaks are only found to thrive when planted out among weeds, such as brambles, which keep patches of soil here and there cool and moist.

The **degree of moisture** contained in any soil is one of the most important factors concerning woodland growth. No plant can grow without imbibing supplies of water; it is requisite for germination; without it, the nutrient salts could not be made soluble for imbibition by the suction-roots, nor could transpiration through the foliage and assimilation possibly take place ; it assists likewise in regulating the temperature of the soil, by making stiff soils milder and loose soils more binding.   But excess of soil-moisture interferes with the normal processes of decomposition, both of the soil and of the humus ; it leads to the development of injurious acids and acid solutions, and tends to the formation of marshes ; seed and roots also suffer from want of oxygen or insufficient aeration of the soil, and hence become easily infected with fungoid diseases ; whilst, in consequence of its poor con-ductivity, and of the amount of heat necessary for raising the temperature of water, it both retards the natural activity of the soil, and increases danger from frost.   Fineness of the soil-particles favours hygroscopicity, or absorption and retention of moisture ; whilst permeability, or diffusive capacity relative to moisture, is proportional to their coarseness.   Soils are in these respects classifiable as follows :—

	Absorptive Power as to		Power of retaining Moisture.	Power of diffusing Moisture. (Permeability.)
	Rain	Aqueous Vapour.		
1	Lime	Clay	Clay	Sand
2	Clay	Lime	Lime	Loam
3	Loam	Loam	Loam	Lime
4	Sand	Sand	Sand	Clay

The addition of humus tends, however, to modify the various characteristic differences exhibited in these respects.

With regard to the degree of moisture they contain, the following classification of soils has been adopted at all the sylvicultural experimental stations in Germany :—

Wet, when water flows out, without pressure being applied, on a clod being lifted up.

Moist, when water falls in drops from a clod on pressure being applied.

Fresh, when only traces of moisture are left on the hand, after pressure being applied to a handful of soil.

Dry, when the soil does not resolve itself into dust on being rubbed. Such soils lose their moisture within a few days after rainfall, and their colour is not much deepened by the moisture retained.

Arid, when, on being rubbed, the soil resolves itself into dust that is carried away by the wind. Traces of rainfall rapidly disappear from such soil, and the slight quantity of moisture it is able to retain does not much affect its colour.

Most of our forest trees—Oak, Beech, Maple, Spruce, Scots Pine, Larch, Silver Fir—thrive best on a fresh soil; Willows, Poplars, Ash, Elm, and Hornbeam prefer a moist soil, and the Alder even a wet one; but stagnating moisture is favourable to no kind of tree-growth. Dryness of soil is not an essential for any of our forest trees; but Birch, Rowan, Aspen, Black Pine, and Scots Pine in general, and Beech on a limy soil, appear to be able to accommodate themselves to less soil-moisture than other species.

The sylvicultural qualities of a soil are frequently, to a very great extent, indicated by the nature of the forest weeds that grow upon it. Thus the following characteristic species of plants may be recognized:

1. *On wet or boggy soil*:—Bog moss (*Sphagnum*), hair moss (*Polytrichum*), cranberry (*Vaccinium Oxycoccos*), bog bilberry (*Vaccinium uliginosum*), heath or bell-heather (*Erica Tetralix*), marsh cistus or marsh rosemary (*Ledum palustre*), cotton grass (*Eriophorum*), sedges (*Carex*), bulrushes (*Scirpus*), rushes (*Juncus*),—the last three genera in many different species.

2. *On fresh fertile land, or soil rich in humus*:—Raspberry (*Rubus idæus*), bramble (*Rubus fruticosus*), red foxglove (*Digitalis purpurea*), willow herb (*Epilobium angustifolium*), deadly nightshade (*Atropa Belladonna*), balsam (*Impatiens Noli-me-tangere*), stinging nettle (*Urtica dioica*), hemp nettle (*Galeopsis Tetrahit*), vetches (*Vicia*), and species of clover (*Trifolium*), as well as ferns and broad-leaved grasses of different sorts.

3. *On drier and more sandy soils*:—Heather or ling (*Calluna vulgaris*), bilberry or whortleberry (*Vaccinium Myrtillus*), red whortleberry or cowberry (*Vaccinium Vitis idea*), whin, gorse, or furze (*Ulex*), broom (*Cytisus scoparius*), greenweed (*Genista*), groundsel and ragwort (*Senecio*),

mullein (*Verbascum*), hawkweed (*Hieracium*), spurge (*Euphorbia*),—these last-named genera occurring in various species,—and the narrow-leaved meadow grasses.

The shrubs which occur most frequently on hills and valleys, especially when the soil is fresh, include the following: alder buckthorn (*Rhamnus Frangula*), blackthorn or sloe (*Prunus spinosa*), hawthorn (*Cratægus oxyacantha*), spindlewood (*Euonymus europæus*), dogwood (*Cornus sanguinea*), barberry (*Berberis vulgaris*), holly (*Ilex Aquifolium*), honey-suckle (*Lonicera Periclymenum*), elderberry (*Sambucus*); on drier soil, juniper (*Juniperus vulgaris*), and on sandy soil, sea-buckthorn (*Hippophaë rhamnoides*).

Relation to warmth is exhibited in changes of soil-temperature in accordance with changes in the atmospheric temperature. It is determined to a far greater extent by the quantity of moisture in the soil, and by its colour, than by its specific warmth. Retentive soils, like clay, are cold and inactive; but, when once heated, they only cool down again gradually. Sandy and gravelly soils become easily warmed and stimulated; but they cool down again rapidly, and hence, in damp localities, increase the danger from frost. In Britain diurnal variations of temperature are obliterated at a depth of about 20 inches from the surface, weekly differences at about 40 inches, and monthly variations at about 6½ feet.

Depth expresses the extent to which decomposition of the soil has taken place below the surface. The finest earth is to be found near the top-level, where the soil has been most exposed to the disintegrating effects of '*weathering*'; the deeper one digs below the surface the larger will the stones and breccia be found, until at last a layer of subsoil is struck which is practically beyond the reach of the decomposing agents, and which may, or may not, be of different geological origin from the soil resting upon it. When decomposition has proceeded so far that the ground is easily penetrable by the root-system for a considerable distance below the surface, a soil is termed *deep*; but when the decomposed and aerated layer of earth is only slight it is said to be *shallow*.

For trees forming a decided tap-root—like the Oak, Scots Pine, and Larch—depth of soil is of very great importance; for, unless the root-systems can develop normally, the growth in height of the trees becomes prejudicially affected. But trees with only moderately deep root-systems, like Beech, Hornbeam, Silver and Douglas Firs, or even rather shallow-rooting kinds like Birch, Aspen, and Spruce, also thrive better on deep than on shallow soils, as these latter generally contain more equable supplies of soil-moisture, and larger stores of food-material in an easily available, soluble form. Deep soils are much less apt to dry up or to become too moist than shallow soils; and the disadvantages of the latter are intensified when the subsoil happens to consist of horizontal layers of stiff clay or other impermeable strata, or of ferruginous deposits like moorpan or limonite. The following is the classification of soils as to depth adopted at the sylvicultural experimental stations throughout Germany:—

Description of Soil.	Depth.	Nature of Woodland Crops for which such Depth suffices.
Very shallow.	up to 6 in. only .	Not really sufficient for any species; but often found under mixed coppice.
Shallow . .	from 6 in. to 1 ft.	Spruce, Aspen, Rowan, Birch, Mountain Pine.
Medium . .	„ 1 to 2 ft. .	Austrian, Corsican, and Weymouth Pines, Beech, Hornbeam, Alder.
Deep . . .	„ 2 to 4 ft. .	Elm, Maple, Sycamore, Ash, Lime, Sweet Chestnut, Silver and Douglas Firs, Scots Pine.
Very deep . .	over 4 ft. . . .	Oak and Larch.

But in Britain, as a matter of fact, woodland soils will seldom be found of great depth: for, when deep, they can generally be used to better economic and financial advantage as arable or pasture land.

IV. Situation in Relation to Woodland Growth.

The growth and the regeneration of the different kinds of forest trees are dependent on combinations of various conditions regarding temperature and moisture, which are united under the term **climate.** Some demand for their normal development an amount of warmth which is absolutely fatal to others ; and these again can bear, and often even prefer, a low temperature during winter, which causes the vital energy of the former to cease, never to re-awaken. Some transpire so freely through the foliage (e.g. Spruce) as to need rather a damp soil and atmosphere, whilst others (e. g. Scots Pine) prefer a drier condition. Without light and warmth the action of chlorophyll in absorbing carbonic acid from the air, and consequently assimilation, would be impossible.

Warmth is in general dependent on, and inversely proportional to, the distance of any locality from the equatorial line, although practically the nature of modifying circumstances connected with the distribution of land and water, high mountain-chains, &c., is always of more or less influence. Thus, in Britain, owing to our insular climate, and to the humid and essentially equalizing tendency of the Gulf-stream, we have no such extremes of summer heat and wintry cold as annually occur in the continental areas lying along the same degrees of latitude in Europe, Asia, or America. Owing to these local modifying influences, there are nowhere hard and fast zones of woodland crops. With ascent above the sea-level, and consequently above the densest and heaviest layers of atmosphere capable of acquiring most warmth in summer, temperature falls at about the rate of $1°$ Fahr. for every 300 ft., and Angot has estimated that for every 333 ft. in vertical ascent woodland growth is retarded for about fourteen days in the awakening of vegetative activity during the spring.

Recent investigations made in the Bavarian Alps (Brenner Pass, 4,500 to 5,000 feet above sea-level) have led to the following conclusions with regard to the effect of vertical elevation on the increment and the shape of forest trees [1] :—

A. *For individual Trees.*
   1. Growth in height diminishes regularly and noticeably.
   2. Growth in basal area, or sectional area of the stem at breast-height, does not diminish so rapidly as growth in height.
   3. Total increment diminishes gradually.
   4. The period of development is prolonged in all of the above three directions (i. e. the attainment of maturity is slower).
   5. The bole or stem deviates more and more from the shape of the cylinder and approximates towards that of the neiloid.
   6. The distribution of the increment over the various portions of the tree increases relatively from above downwards.
   7. The form-factor at breast-height (i. e. the proportion which the actual volume of the stem bears to a cylinder having the same base) becomes smaller.
   8. The coronal development gradually comes lower down, nearer to the ground (i. e. the stems become short and stunted).
   9. The proportion of branches and brushwood increases (i. e. as the stems grow stunted, the coronal development increases relatively).

B. *For whole Crops.*
   1. The number of stems per acre increases, but the proportion of those which form the larger girth-classes diminishes.
   2. The mean average height of the crops diminishes.
   3. The total basal area at breast-height does not diminish so perceptibly as the average height of the crops, but it is made up mostly by units belonging to the smaller classes of stems.
   4. The total quantity of wood available as timber or fuel decreases perceptibly ; hence the total average annual increment of the mature crop becomes considerably less.
   5. The quantity of small branches and brushwood increases.
   6. There is a decided tendency towards growth of the trees in groups instead of being equally distributed over the whole area.

Near the polar regions, or at very high elevations, there is no woodland growth.   This may not be solely due to the actual

---

[1] S. Honda, *Ueber den Einfluss der Höhenlage der Gebirge auf die Veränderung des Zuwachses der Waldbäume*, 1892, p. 27.

amount of cold in winter. It may, as Hartig has shown [1], be caused by transpiration being excited through the leaves of the evergreen conifers (which reach to the highest latitudes and elevations) on bright, sunny days, whilst the soil still remains frost-bound and unable to allow of imbibition of fresh supplies of moisture by the root-system in order to replace what is being evaporated. Wilting and death are then the inevitable results if the winters are long and dry, and if the sunny days have been frequent.

But, although woody plants suffer comparatively little from the effects of winter cold, yet late frosts in spring are apt to injure the young leaves and shoots, and also early frosts in autumn before the shoots have hardened. These dangers are greater in low-lying and confined localities than where aerial currents have free play. Ash, Acacia, sweet Chestnut, and Beech are most sensitive to frost; Lime, Hornbeam, Elm, Birch, Larch, Aspen, Austrian, Corsican, and Scots Pines are decidedly hardy; whilst Oak, Silver and Douglas Firs, Maple, Sycamore, Spruce and Alder occupy an intermediate position, in which the influence of other factors than atmospheric temperature merely (e.g. humidity of soil and air) determine whether they are likely to incur danger or not.

The absolute equivalent of heat necessary for the normal development of any given species of tree is as yet unknown. But experience shows [2] that Elm, sweet Chestnut, pedunculate Oak, and Black Pines require the greatest amount of warmth, and that Larch, Cembran, and Mountain Pines can do with least; whilst, between these extremes, Silver and Douglas Firs, Beech, sessile Oak, Scots and Weymouth Pines require decidedly more than Maple, Sycamore, Ash, Alder, Birch, and Spruce. Although their general requirements as to warmth limit the different species to more or less normal zones of

---

[1] *Lehrbuch der Baumkrankheiten*, 2nd edit. 1889, pp. 104, 261. But compare the remarks already made on p. 68.

[2] Gayer, *op. cit.*, p. 20.

elevation, yet quality of the soil, aspect, and local climatic conditions modify these demands in the most marked degree. It is more particularly the amount of warmth during the period of active vegetation which determines whether or not the different species of trees can thrive and regenerate themselves naturally in any given locality. If, between the time of their flowering and the natural term of their ripening the fruit, there has not been a sufficient development of warmth, accompanied by an absence of destructive frost, to enable the seed to attain maturity and germinative capacity, then the natural conditions for thriving and regeneration are not offered to any given kind of tree. The further northwards deciduous broad-leaved trees occur, the later is usually the time at which they break into leaf, as the amount of warmth requisite for stimulating active vegetation is only obtainable later in spring than at similar elevations situated further southwards.

Light is also a factor which is undoubtedly of immense importance with regard to the thriving of woodland crops. A careful study of their demands with respect to light, and their capacity for enduring shade, will quite logically explain many of the concrete conditions in which woodland growth exhibits itself, as was first pointed out by Gustav Heyer[1] (see also p. 54).

The **Relative Humidity** of the atmosphere is governed by the atmospheric temperature ; for, as regards any given quantity of moisture, the saturation-point will be much sooner reached when the air has only a low temperature, than when its temperature is high[2]. But, whilst the absolute humidity of the air varies inversely to the latitude above the equator, no such rule obtains regarding its relative humidity, as local circumstances determine the supplies of moisture available for evaporation. Thus, in Britain, the soft, mild, westerly winds which reach this island only after sweeping over the Atlantic

---

[1] *Das Verhalten der Waldbäume gegen Licht und Schatten,* 1852.
[2] See Roscoe's *Elementary Text-book of Chemistry,* 1888, pp. 55, 56.

in the track of the Gulf-stream, have a high relative humidity; whilst those from the east, which come from about the central area of Asia and Europe, are comparatively dry, although (in consequence of the moisture acquired in passing over the North Sea) not so dry as on the continent, where the relative humidity of the air decreases with longitudinal progress eastwards.

As aqueous precipitations must occur whenever the air is reduced so far in temperature that the corresponding saturation-point renders it impossible for all the moisture to be retained, rainfall is more frequent at higher elevations than at lower levels, whilst at the same time aerial disturbances of various kinds are also more frequent and violent.

For the maintenance of normal transpiration through the foliage of woodland crops enormous supplies of moisture are requisite in the soil. When a period of drought occasionally sets in during summer for about a fortnight to three weeks, this often forces vegetation to close its activity for the year in the case of broad-leaved trees, or can even be fatal to species like the Spruce, which transpire freely[1]. Though the actual amounts of soil-moisture requisite for the various species of trees of woodland growth are not accurately known, yet observation and experience show that Spruce, Black Pines, Alder, Maple, Sycamore, and Ash thrive best in a humid atmosphere, whilst Beech, Birch, and Silver Fir grow better in damp than in dry localities. If, therefore, the former species be planted out in situations where the air is usually dry, they make considerably greater demands as to soil-moisture than when growing in localities where, owing to a higher relative humidity of the atmosphere, the transpiration of water through the foliage is not so much stimulated[2].

The **Aspect, or Exposure** towards one or other of the cardinal points of the compass, exerts great influence on the

---

[1] Compare Chapter VIII, Section 1.
[2] See *British Forest Trees*, 1893, pp. 27, 28.

temperature of soils, and on the amount of moisture retained in them. Southern and south-western aspects, owing to their greater warmth and to the stimulation of evaporation, both through the foliage and direct from the surface-soil, are usually much drier and more apt to get heated than northern and north-eastern exposures, which are cooler and damper, and whose soil is usually less active both in awakening vegetation in spring, in stimulating to assimilative processes, and in decomposing the dead foliage on the ground so as to form humus. On high ranges of mountains, however, trees will still be found on the warmer aspects at levels above those limiting their growth on the northern slopes. Eastern exposures are most liable to danger from late frosts; whilst western and south-western aspects are most exposed to damage from wind-fall and breakage, as the heavy storms coming from these directions are usually accompanied by rain, and often occur at times when the soil is still sodden from recent downpours.

The **Slope or Gradient** of any soil is necessarily of influence; for the opportunities which rainwater has of percolating down to the lower layers decreases as the angle formed with the horizontal by the surface of the soil increases. According to Grebe, soils with a slope of from 5 to 30° form the true home of woodland growth; hence much land which is too steep for agricultural cultivation is well adapted for sylvicultural occupation. Grebe's classification of gradients, now adopted at the experimental stations throughout Germany, is as follows :—

gentle . . . . 5–10°	*steep* . . . . 21–30°
medium . . . . 11–20°	*very steep* . . . . . . 31–45°

precipitous over 45° ; this is, however, incapable of sustaining tree-growth.

The **Configuration of the Soil and of the surrounding Country** exerts a considerable amount of influence on the woodlands. The soil near the base of hills is generally deeper, richer, and more productive than on the slopes or ridges ; but the danger from frosts is greater in hollows and throughout low-lying tracts than where currents of air have free play.

Mountain masses have greater uniformity of soil and situation than ranges of hills. On these the well-being of the woodland growth depends to a great extent on the amount of protection that can be given to it from winds; for the bad effects of want of protection may only too clearly be seen in exposed plantations or woods near the sea-coast or on bleak hill-sides.

In the assessment of the quality of soil the best standard is that which makes five distinctions, thus :—

I. *Very good*; II. *good*; III. *average or moderate*; IV. *inferior*; V. *poor*.

In Germany the decimal system is often adopted. But practical men will have less difficulty in making up their minds as to whether any given soil is average or inferior in quality, than they will find in estimating it at 0·3, or 0·4, or 0·5 of the first-class soil classed as equivalent to 1·0. Such assessments are very useful, but are in each case only relative; for a soil that might only be regarded as inferior so far as the cultivation of mixed crops of Oak, Maple, Sycamore, Larch, Douglas Fir, and Beech are concerned, might perhaps be considered a moderately good, or even a good soil if considered with regard to woods in which Pines, Spruces, and Firs were to be the ruling kinds of trees.

CHAPTER VI

ON THE ADVANTAGES OF MIXED TIMBER-CROPS OVER PURE WOODS

AT p. 33 of the *Nineteenth Century* for July, 1891, Sir Herbert Maxwell, Bart., in his article on *Woodlands*, in endeavouring to assist landowners towards a better system of Forestry than at present exists throughout Britain, makes a statement that the general bad growth of our forest trees *as timber trees* is attributable to the fact that—

'Mixed planting is generally practised, in sharp contrast to what Continental foresters call " pure forest "—that is, a woodland composed of one species of tree. This is in itself a hindrance to profitable management, because pure forest is much more easily tended than mixed plantation, and the timber is more readily marketable.'

In England we are not accustomed to look upon the French as a particularly practical nation ; but, in regard to the treatment of their great forests, with which 17·7 % of the total area of France is clothed, it must be admitted that, although the methods adopted are not quite so scientific as in Germany, they are generally of a very high standard and thoroughly practical. Now in France, where sylviculture is well understood and practised, mixed forests form the great bulk of the woodlands, as, according to the last statistics available (Mathieu's, for

1876) these formed more than 70 % of the total wooded area, as exhibited in the following data [1] :—

Mixed forests of—
    Broad-leaved trees   50·3 % of total wooded area.
    Broad-leaved trees and Conifers 17·6 %   ,,   ,,
    Conifers   2·5 %   ,,   ,,
Pure forests of —
    Various species of Trees   26·7 %   ,,   ,,

Data for Germany are not available. The imperial statistics for 1884 only showed 65 % of the total forest area (which aggregates 26 % of the empire) as covered by coniferous crops, whilst 34·5 % were clothed with broad-leaved species; but the area of mixed forests was not computed.

In sharp contrast to Sir Herbert Maxwell's dictum, is the following matured opinion of the most celebrated of living sylviculturists, Prof. Gayer of Munich [2] : —

'We may say, in general, that the leading principle in the rational and economic treatment of woods must be less in the direction of pure than of mixed crops, and that the degree to which, and manner in which, mixed woods occur throughout any system of management must be considered as the best *test and standard by which one can estimate the knowledge and the capacity* of those to whom are entrusted the duties of obtaining the best possible results from any given conditions as to soil and situation.'

And, to prove that this is no mere new nostrum of these latest days, it may be permitted to quote the late Prof. Grebe of Eisenach, formerly Director of the Saxon forests in Thuringia, who wrote as follows in 1867 [3] :—

'The more recent tendency of forestry is—and with full justification—towards the formation of high timber forests containing a *suitable mixture of different species of trees*, thereby duly acknowledging the manifold advantages to be gained with respect to increased outturn, more valuable assortments of timber, greater general security, and greater power of resisting inimical influences.'

---

[1] Gayer, *Der gemischte Wald*, 1886, p. 9.
[2] *Der Waldbau*, 3rd edit., 1889, p. 178.
[3] *Die Betriebs- und Ertragsregelung der Forste*, 1867, p. 151 (2nd edit., 1879). See also Preface to the author's *British Forest Trees*, 1893.

As arboriculturists we have nothing to learn from continental nations ; but as sylviculturists we may learn a very great deal indeed.   There is a need for the dissemination of sound knowledge, in the first instance regarding the proper density at which sowings and plantations should be made, and in the second instance as to that which should be maintained afterwards at all the various stages of their development.   Sir Herbert Maxwell attributes the formation of mixed woods in great measure to neglect in cutting out the nurses originally planted for the purpose of stimulating the crops to quicker growth and of keeping down weeds.   He says—and the statement is undoubtedly a true indictment of neglect and incapacity in those charged with the supervision of woodlands :—

'Want of system leads to irregularity in thinning out the nurses, which often remain to compete with what was intended to be the permanent wood, and the result is a mixed plantation.'

As will be pointed out in another chapter (see p. 183), such nurses should be removed during the *weedings and clearings* of the young woods ; whilst, if for any reason they have been allowed to stand till *thinnings* commence, an opportunity is then still given to remove them gradually.

It would, however, hardly be rational to set up these mismanaged woods—admittedly the outcome of neglected tending operations, and not at all in conformity with the intentions or wishes of the proprietor—on anything like the same platform as mixed woods formed after deliberate consideration as to the best kinds of trees suited for each portion of the soil to be utilized under woodland crops.   But, where mixed woods are formed by capable foresters, they have, in comparison with pure forests, the great advantages of denser growth, larger and finer production of timber both as regards quantity and technical quality, comparative immunity from throwing or breakage by wind, snow, or ice, as well as from dangers arising from insects, fungoid diseases, and fire ; whilst against all these very solid advantages only one drawback can be named,—that the tending of such mixed woods is much more difficult, and that their proper formation and management require considerably greater knowledge of sylviculture than is requisite for the treatment of pure forests.

In comparing the advantages and disadvantages of mixed woods, relative to pure crops, the following matters are worthy of consideration :—

Advantages—

1. A greater density of crop is obtainable.

2. The soil is maintained in better condition as to productive capacity than under many pure crops.

3. A larger percentage of the outturn is available as timber for the higher technical purposes, at any rate in mixed crops of broad-leaved species of trees.

4. Demands of varying nature for timber can more easily be supplied.

5. It is easier to modify or transform the crop at any one time so as to meet the present or the probable future requirements of the market.

6. Experience has shown that mixed crops are much less exposed than pure forests to dangers from external causes whether of organic or inorganic origin.

7. Natural regeneration of mixed woods is on the whole easier than with pure crops.

8. By concentration of various species in one series of crops the operations of tending, harvesting, &c., can be carried out most economically.

Disadvantages are summed up by Gayer in the following words [1] : —

'It is easy to understand that mixed crops make higher demands on the capacity of the sylviculturist, as it is much more troublesome and difficult to give proper care and attention to each of several species in

[1] *Waldbau*, p. 216.

a mixed crop than to look after one species only; but when weighed against the many solid advantages offered by mixed crops, this drawback can surely seldom be seriously taken into account.'

Mixed woods are such as are formed by two or more species, of which the matrix or ruling species should be one that is capable of safeguarding and retaining the productive capacity of the soil against insolation and the exhausting action of winds. The subordinate or dependent species of trees may be admixed with the ruling species either in *clumps* or *hursts*[1], in *groups* or *clusters*, in *patches* or *knots*, in *rows* or *lines*, or merely scattered about *singly* as individuals. In mixed woods, wherever natural regeneration is permitted to the fullest extent, there will be a tendency for genera of trees with heavy seeds (Oak, Beech, and among Conifers the Silver Fir) to reproduce themselves in clumps on the patches of soil best suited to them; whilst trees with lighter and often winged seeds (like Birch, Aspen, and Willow) will have a much greater repro-ductive power of asserting themselves individually and in small knots at some distance from the parent trees.

1. *A greater Density of Crop is obtainable.*—Wherever the area under woodlands is of any considerable extent there are almost certain to be many variations of soil and situation with regard to the nature and the depth of the earth, its degree of moisture, its aspect, &c.; and the best, most complete, and most economical utilization of the soil can only be expected when each such varying portion is stocked with the species of tree best suited for growing there. For, when thus growing in admixture on the patches of soil most suitable to their natural requirements, it will be found that the various species of trees do not thin themselves so soon as they otherwise must on soils less favourable to their natural development and requirements. And, by thus maintaining close canopy for the longest possible period, they not only avail themselves more fully of the pro-

---

[1] The older term *Bosc*, often referred to by Manwood (*History of the Forest Laws*), is now obsolete.

ductive capacity of the soil, but also conserve this in a higher degree than could otherwise be the case; whilst, at the same time, they are able to utilize to the utmost possible extent the chemical power of the light requisite for the due performance of the assimilative functions by the foliage. This closer canopy and greater density of crop generally become of all the more importance as the crops advance in age, and begin to approach maturity; hence in this prolonged maintenance of canopy lies the principal value of many kinds of mixed crops.

2. *The Soil is maintained in better Condition and Productive Capacity than is the case under many kinds of Pure Crops.* Pure woods of thinly-foliaged and light-demanding genera of trees, like Oak, Ash, Maple, Larch, and Pine, when once they have completed their chief growth in height and have begun to exhibit their natural specific tendencies with regard to increase of growing-space, are unable to conserve for themselves the general quality and productive capacity of the soil in consequence of the interruption of canopy, the decrease in the number of stems forming the crop, and the subsequent bad effects to the surface-soil wrought by insolation and exhausting winds. Hence the expenses of underplanting are usually unavoidable. But as, during the earlier periods of development, these light-demanding genera are at the same time of more rapid growth than the shade-bearing kinds, Beech, Spruces, Silver Firs, and Hornbeam, in mixed crops they can, with now and again a little assistance in the way of tending wherever necessary, usually assert themselves as the predominating poles and trees, thereby acquiring just the position naturally best suited for their present growth and good future development. At the same time the shade-bearing genera, forming the dominant and more or less dominated classes of the crop, mechanically protect the soil against the exhausting and deteriorating influences of sun and wind; whilst, by their richer fall of leaves, they contribute valuable material for the formation of humus or mould, that natural manure of woodlands, which exerts very

beneficial influence in enhancing the productive capacity of most classes of woodland soil.

Pure forests make constant demands of one unvarying sort on the land, so that on soils of merely average quality there is far greater risk of one particular class of nutrients being rapidly diminished and being less readily available than when a variety of species of trees induces a more general demand for the various different kinds of mineral food-substances. Thus, whilst in general the chemical composition of dry woody-substance of the different kinds of timber is, in round figures, approximately constant as follows [1] :—

50 % *Carbon*, 42 % *Oxygen*, 6 % *Hydrogen*, 1 % *Nitrogen*, and 1 % *Ash* (or mineral residuum),

yet the nature of the substances found in the ash varies greatly in the different species of trees. Thus, with regard to lime, the principal constituent in the ash, there are 3–5 times as much found in Oak, and 2–3 times as much in Beech, as in the Scots Pine, which, however, exhibits more than Birch. Taking the quantity contained in Scots Pine as unity, then the amount of potash requisite is about 5 times greater in Beech, $3\frac{1}{2}$ times greater in Oak, $2\frac{1}{2}$ times greater in Silver Fir, twice as much in Larch and Birch, and about $1\frac{3}{4}$ times as much in Spruce. And again, with regard to Phosphoric Acid, the wood of Beech contains in its ash about $2\frac{1}{2}$ times as much as Scots Pine, Oak 3 times as much, Birch twice as much, Larch and Silver Fir $1\frac{1}{2}$ times as much, whilst Spruce contains considerably less.

There can be little doubt that the theory enunciated by Georg Ludwig Hartig, the father of Dr. Theodor Hartig, and grandfather of Prof. Robert Hartig (of Munich), that the soil can be most thoroughly utilized by an admixture of shallow-rooting species with deep-rooting, in order that each may draw

---

[1] Weber, *Die Aufgaben der Forstwirthschaft* (Lorey's *Handbuch*, &c.), 1886, vol. i. pp. 72 and 62. More accurate details have already been given in Chapter IV. pp. 82–85.

its main supplies of nutriment from a different layer of soil—
although long scoffed at by a certain class of sylviculturists
(headed by Hundeshagen), who saw the advisability of admixture explained solely by the relation of the various species
towards light and shade—is based on a principle equally sound
from the theoretical and the practical points of view; for it is
a fact that, in the majority of mixtures acknowledged to be good,
the root-systems of the different species vary considerably in
depth.

That, in addition to greater density of growth, which has
already been referred to in the previous section, the productive
capacity of the soil is at the same time more thoroughly utilized
can be proved, if requisite, by actual crop measurements made
in Silesia in 1880.   These showed that, on soil similar as
to conditions and quality, the average annual increment in
eighty-year-old crops was :—

Cubic ft. per acre.

Pure forest of Scots Pine	18.3
Pure forest of Spruce	19.9
Mixed forest of Scots Pine, Spruce, and Silver Fir .	23.5

But, beneficial though the influence of the overshadowing of
the soil by the leaf-canopy undoubtedly be, there are occasions
on which even it may perhaps be carried farther than is desirable or advantageous; as, for example, in dense woods of
pure Spruce or Silver Fir, where the close, thickly-foliaged
canopy overhead, and the deep sponge-like growth of moss
covering the soil, intercept and retain a very much larger proportion of the atmospheric precipitations, both in winter and
in summer, than deciduous trees or other coniferous species
(Pines) with somewhat thinner growth.   For whilst, of all the
precipitations in the course of the year, on the average about
25 % are intercepted by the foliage and branches, and 75 %
reach the soil in woodlands, these data vary considerably
according to the kind of tree forming the crop.   The averages

of long-continued observations in Switzerland, Prussia and
Bavaria gave the following results [1] :—

Species of Tree.		Percentage of total Atmospheric Precipitations	
		reaching the soil.	intercepted by, and evaporated from, the foliage and branches.
Deciduous	Larch . .	85	15
	Beech . .	81	19
Evergreen	Spruce . .	76	24
	Scots Pine	70	30

And of the moisture reaching the soil, Weber calculated that—
leaving out of account the transpiration of the foliage for which
the supplies of moisture are withdrawn by the roots from deeper
layers of soil—its disposal was as follows :—

Woods.	Amount evaporated.	Remained in the soil.
	per cent.	per cent.
Beech	40·4	59·6
Spruce	45·3	54·7
Scots Pine . . .	41·8	58·2

Owing to the smaller percentage of precipitations reaching
the ground, to the larger evaporation from the soil-covering, and
to the strongly hygroscopic nature of the moss, we can easily
understand the dried-up appearance that soils under pure
Spruce woods sometimes assume in early summer, just at the
time when transpiration is becoming most active, and when the
greatest necessity exists for a free supply of moisture from the
soil so as to permit of the assimilative and productive processes
going on to their fullest possible extent.

In former times, when the work of nature was comparatively
uninterfered with, mixed woods almost everywhere formed the

---

[1] Compiled from data given by Weber, *op. cit.*, pp. 47–49.

covering of the uplands, and the hilly tracts, and stretches not yet required for arable land or pasturage.   Rain and snow were then more equably precipitated over the soil itself, and were not so apt to be sucked up and retained by a strongly hygro-scopic surface-growth of weeds and mosses, which likewise is exceedingly active in afterwards utilizing and exhausting the supplies of moisture contained in the upper layer of the soil. The aqueous precipitations were, on the other hand, preserved by a good layer of dead foliage and of the humus or mould resulting from their decomposition, and could thus percolate normally into the soil downwards towards the subsoil, so as to maintain a tolerably equable distribution throughout the whole, owing to the effects of gravity being counterbalanced by capillarity.

Whilst Boussingault has shown in France that, with increase in the area stocked with conifers, a gradual local sinking of the water-level below the soil has taken place, practical sylvi-cultural experience in Germany has also shown that—except at high elevations with humid climate—extensive plantations of pure Spruce rapidly lead to a marked decrease in the quantity of soil-moisture.   And Runnebaum's comparative investigations in 1885 showed that in mixed woods of Scots Pine with Beech the conditions relative to soil-moisture were more favourable for tree-growth than in pure forests of Pine. It is, however, only right to point out that this must to a certain degree be ascribed to the influence exerted on the soil by the good humus or mould formed by the dead foliage of the Beech.

3. *A larger percentage of Timber is available for the higher technical purposes, at any rate in mixed crops of broad-leaved genera of trees.*   Many of our more valuable kinds of timber trees are, when once they have passed through the initial stage of pole-forest, so light-demanding, and so dependent on increase of growing-space for the continuation of vigorous growth and further normal development, that they are unable to maintain

for themselves such degree of density of canopy as is requisite
for the due conservation of the productive capacity of the
soil, and for the prevention of its gradual, but inevitable,
deterioration.

Even on the better classes of soil, pure crops or mixed woods
of valuable light-demanding species only—Oak, Ash, Maple,
Larch, Pine, &c.—grown without admixture of any shade-
bearing species, are apt to thin themselves out naturally at so
rapid a rate as to favour branch and coronal development to
an extent inconsistent with a studiously economical utilization
of the woodland soil. This, too, is a drawback not in any way
obviated by underplanting. But an admixture of one or other
of the shade-bearing species (Beech, Hornbeam, Spruces, Silver
Firs), according to the nature of the soil and situation, not only
protects and stimulates the productive capacity of the soil, but
at the same time acts mechanically in stimulating to growth in
height, and in preventing the formation of short boles and of
stout branches that would prejudice the technical utility and
the market value of the timber produced.

It is a matter of practical experience, needing no discussion,
that Larch and Scots Pine form more valuable boles, and
contain a larger percentage of heart-wood, when grown in
admixture with Spruce or Silver Fir than as pure crops.
And all the more valuable species of deciduous trees gain
in shape, increment, and age up to which they may be advantageously grown from technical and financial points of view,
when the Beech, or on moister descriptions of soil, the
Horn-beam, forms a part of the crop. But these beneficial
effects are even more distinctly noticeable when mixed crops
are formed, on suitable soils, of conifers and broad-leaved
trees—even when all are shade-bearing, as in the case of
Spruce and Silver Fir forming crops along with Beech.

In consequence of the greater increase in the total quantity of
timber produced in mixed crops as compared with pure woods,
and of the more favourable circumstances affecting the forma-

tion of long, full-wooded, clean stems with comparative freedom from knots and branches, the timber produced attains its highest possible technical and financial value. And owing to the tendency towards confinement of the crown to the upper part of the tree only, to which these favourable results are practically due, a considerable portion of the upper wood is comprised within the bole that must otherwise have been dissipated in the formation of branches of comparatively little technical or monetary value. By concentrating the energy of growth in good bole-formation, the actual cubic content of the stem is not only increased, but also its value per cubic foot, as this rises with the length, straightness, and freedom from flaws occasioned by knots and branches.

4. *Demands of varying nature for Timber can more easily be satisfied.* This statement may almost be said to be of an axiomatic nature. It plainly stands to reason that, where several species of trees are grown together on the same area, the classes of material periodically yielded during the necessary operations of tending, thinning, and removal of diseased or already mature individual stems, must offer a considerably greater variety to the different classes of local consumers than is possible in the case of pure forests. And the same holds good when the period comes for the harvesting of the mature crop. At the same time that variations in the local demands can be more easily satisfied, the wider, more general, and more important requirements of timber marts at some distance can also be better met by mixed than by pure forests; whilst there is less danger of the over-production of any one kind or assortment of timber to any such extent as might possibly glut the market and cause a sudden fall in the prices obtainable.

5. *It is easier to modify or transform the Crop at any time so as to meet the present or the probable future requirements of the market.* Of the necessity for such considerations no better example can be given than is shown in the Crown Oak woods of the New Forest, Alice Holt, Parkhurst, and the Forest of

Dean, which were planted up for the purpose of supplying Oak timber for the use of our navy. But now all our ships of war, and the best classes of liners, are constructed almost entirely of iron or steel and of Teak timber from Burma. Teak—in place of corroding iron with which it comes in contact, as Oak does owing to its tannic acid—is, through an essential oil contained in it, preservative of the steel screws, bolts, &c. But notwithstanding the present extensive and ever-increasing use of substitutes like iron in place of timber, and the vast improvements in communication—by means of which the forest produce of far-distant countries can be brought to build the navies of Europe, and even to pave the streets of London—timber is, along with gold, almost the only commodity which has not during the past half century depreciated in relative exchangeable value with most other articles of commerce. All classes of timber do not, however, command equally favourable market quotations. But the sagacious sylviculturist will easily know from a study of selling prices, of the stocks of timber held at the chief marts, and of the reports as to the probable quantity and the quality of imports from abroad, what kinds and classes of timber are most likely to be remunerative in the immediate future. And it is self-apparent that he will be in a far better position to tend and foster those species which hold out the fairest financial promise, when several kinds of trees are growing together in mixed woods, than when each species forms a pure forest worked quite independently of the other crops.

Mention has above been made of the Oak woods planted by the Crown for the future supply of navy timber, which is not now required. As crooks were formerly an essential before the framework of hulls was made of iron or steel, the most economical method of producing them was to grow the Oaks as standards enjoying as full supplies of light and air as possible. But, under ordinary circumstances, the most economical manner of growing Oak timber for general technical purposes (requiring length and full-woodedness of bole, straightness, and freedom from knots or twists), and for the highest possible financial returns, is undoubtedly to confine the individual stems within the least growing-space sufficing for their actual requirements as to light, warmth, and the normal activity of their assimilative organs, until the main growth in height has been completed, and until natural demands for freer exposure to light and air, and greater lateral expansion of the crown of foliage, can no longer profitably remain unheeded. Crops, however, which have been originally formed and treated with the express view of yielding crooked stems, cannot be forced later on to form fine, straight, clean boles showing a minimum of tendency to tapering growth near the top-end. Such existing crops must first be cleared and regenerated before others can be formed and produced with the express intention of utilizing the productive capacity of the soil in the fullest and most economical manner.

And, on the other hand, whenever present market rates offer inducements for the speedy clearance and sale of certain classes of timber, advantage can much more easily and conveniently be taken of the favourable opportunities of disposal without at the same time threatening the well-being of the woodlands by departing to a greater or less extent from the normal methods of treatment that may have been forecast by the provisions of the Working Plan. And a Working Plan must be the basis of all operations in any well-managed forests.

In short, when the growing-stock consists of one series of falls of mixed woods, varying in age from zero to maturity, in place of having a similar series for each of the species of trees grown by themselves in pure crops, the Working Plan is capable of much greater elasticity of treatment. It affords larger scope for the somewhat premature utilization of portions of crops, whenever the market may be exceptionally favourable for that particular kind or assortment of timber, without necessarily exposing the soil to deteriorating influences likely to weaken its productive capacity. By removing or diminishing the number of individuals of other species during the ordinary operations of thinning and tending, or, if necessary, even in special revisions for this purpose, the probable immediate future requirements of the timber marts within reach can be more economically, conveniently, and speedily arranged for than when one has to deal with pure forests only.

As an example of this the safety-match industry throughout Norway, Sweden, and Germany, may be instanced. The woods best suited for the requirements of this industry are Willows, Aspen, and Poplars; and it very frequently occurs that the match-factories have to be removed from one district to another solely on account of the supplies of suitable timber falling off. The heavy expenses thus involved mean that if steps were taken to provide better supplies of these softwoods, the factory proprietors could well afford to pay higher rates to secure them. The full advantage of such favourable markets can only be derived when patches of moist soil have been utilized for the growth of these softwoods in mixed deciduous forests.

6. *Experience has shown that Mixed Woods are much less exposed than Pure Forests to dangers from external causes, whether of organic or inorganic origin.* Shallow-rooting species, like the Spruce in particular, when mixed with deeper-rooting kinds of trees, like Pine, Larch, and Silver Fir, are much less exposed to be damaged or thrown by storms than when grown in pure crops; whilst conifers of all kinds suffer less damage from wind, fire, snow, ice, insects, or fungoid diseases, when grown in admixture with broad-leaved trees than if forming pure woods, or even mixed woods consisting of conifers only.

If associated along with broad-leaved trees, conifers usually attain a better development than under any other circumstances. This alone would partly account for the decreased liability to attacks from insect enemies; whilst, at the same time, a much better check is kept on any tendency to increase in the number of insects, as insectivorous birds are much more frequent where broad-leaved trees exist than in purely coniferous woods.   But even where attacks of noxious insects take place in such mixed woods, they seldom become widespread over all species, and never necessitate the clearance of whole crops, as unfortunately sometimes happens in the case of Spruce and Scots Pine.

Trees like the Oak, which are somewhat sensitive to late frosts in spring during the early period of their growth, are apt to suffer less damage whilst growing under the protection of hardier kinds of quicker development, like Birch and Scots Pine.   And again, in mature mixed woods of conifers and broad-leaved species, the technical value of the boles is less likely to be diminished by frost-shakes than when these latter are grown by themselves.

The woodland proprietors, of Scotland especially, know what ravages can be committed in pure Larch forests by the fungoid disease, due to *Peziza Willkommii*, which causes a cankerous outbreak on the stems, and induces crooked growth.   And in pure woods of Spruce, Silver Fir, and Scots Pine, red-rot and other fungoid diseases of the stem and the root-system, occasioned by *Trametes radiciperda*, *T. pini*, *Aecidium elatinum*, *Peridermium pini*, *Agaricus melleus*, &c., are often productive of very serious sylvicultural and financial results.   Speaking of '*leaf-shedding*,' a disorder to which Scots Pine is liable at an early age—an infant ailment, due sometimes to frost, or to the drying-up of the needles when stimulated to transpiration by bright sunshine in winter or early spring, whilst the soil is still frost-bound and unable to yield fresh supplies of moisture to replace that evaporated, but in many other cases undoubtedly occasioned by a fungus, *Hysterium pinastri*,—Gayer says[1] :—

'About thirty years ago the sporadic occurrence of "*leaf-shedding*" in Scots Pine was merely "*an interesting observation*," whereas now it is

---

[1] *Der gemischte Wald*, 1885, p. 24.

nothing short of a wide-spread calamity all over Germany. At first, two
to four-year-old seedling growth was chiefly attacked, but nowadays
sowings and plantations are both liable to be affected, and it is even beginning to attack young thickets. And wherewith shall we replace the
Scots Pine, if even it refuse to serve our purposes? R. Hartig says, at
page 40 of his book on the Diseases of Trees[1]:—"*the best prophylactic
measure against the occurrence and spread of epidemics is the cultivation
of mixed forest crops.*"'

A direct and fuller quotation may, however, also be permitted from Professor R. Hartig's celebrated work, in which
the opening words of the Introduction are :—

'The transformation in the natural woods of Germany, the formation
of pure. equal-aged crops of the same species of trees, instead of the
composite mixed forests and copses made up of all sorts of trees of all
ages, and especially the restriction of the broad-leaved species by the
formation of pure crops of conifers, have during the present century,
and most particularly in the last few decades, threatened the well-being
of our woods to an extent which was formerly unknown. And it is
chiefly the enemies belonging to the animal and vegetable kingdoms
which find in the recent developments of our woodland economy the
conditions favourable to their increase in enormous strength, so that the
complaints made concerning larger and ever larger devastations, appear
in no wise unfounded.'

But it is perhaps more with regard to damage caused by
injurious insects, than in any other way, that the beneficial
effects of the formation of mixed crops may be judged of, so
far as Spruce and Scots Pine are concerned. Simultaneously
with the disappearance of the natural mixed crops, which can
be proved to have formed the chief portion of the wooded
areas throughout Germany during the first quarter of the
present century, and with the formation of extensive pure
forests of coniferous species, the damage done by insect
enemies has increased enormously, and has on several occasions during the past half-century attained the dimensions of
actual calamities, sometimes of national importance.

Thus, referring to the ravages of the Black Arches, ' Nun,'

---

[1] *Lehrbuch der Baumkrankheiten*, 1882 (2nd edit. 1889, p. 55).

or Spruce moth (*Liparis monacha*) which—together with the bark-beetles (*Bostrychini* and *Hylesinini*) that followed it and attacked the stems left in a sickly condition of growth—between 1853–1862 devastated the coniferous forests of eastern Prussia and western Russia over about 7,000 geographical square miles, and necessitated the felling of over 420,000,000 cubic feet of timber, Hess recommends[1], as the first measure of prevention, '*avoidance of the formation of pure forests of Spruce or Scots Pine; formation of mixed crops of suitable species.*' This same insect has quite recently, in southern Bavaria and western Austria (1889–1891), necessitated the clearance of 40,000 acres, mostly of pure Spruce; and during 1892 an outlay of £75,000 was incurred solely in order to combat its further attacks in the State Forests of Bavaria alone[2].

That the calamitous extent to which attacks by moths (*Bombycidae*), bark-beetles (*Scolytidae*), and weevils (*Curculionidae*) occur every now and again in Germany, arises from the want of natural admixture of different species is a fact which receives practical recognition in the endeavours now everywhere being made in order to effect a re-transformation to the former mixed state of the woodlands. As Gayer says[3]:—

'Endeavours are in general being made to prepare the way for a return to mixed forests in all suitable localities. And that, face to face with the late fearful devastations (by insects) in the Spruce forests of southern Bavaria, these principles should be even more strongly insisted on, can easily be understood.'

It is a well-known matter of experience that forests of broad-leaved species are not exposed in anything like the same degree to serious damage from insects. Even after crops may have been stripped of their spring flush of leaves—as sometimes

---

[1] *Der Forstschutz*, vol. i., 1887, p. 355.
[2] A detailed account of this moth and its ravages will be found in the *Transactions of the Highland and Agricultural Society of Scotland*, 1893, pp. 176–207.
[3] *Zeitschrift für Forst- und Jagdwesen*, 1892, p. 386.

happens when swarms of caterpillars of the Processionary Moth (*Gastropacha processionea*) defoliate the Oak, for instance—the injury is not of a permanent nature, owing to the much larger reserves of starchy matter accumulated by the broad-leaved kinds of trees, and the stronger reproductive and recuperative capacity with which they are consequently endowed.

By admixture of broad-leaved genera, therefore, and particularly of Beech or Hornbeam, along with conifers, the danger of attacks from insects can be greatly diminished without prejudicing the other advantages derivable from the formation of mixed crops. Thus Danckelmann states [1] that woods which consisted of four-fifths Scots Pine and one-fifth Beech or Hornbeam practically did not suffer at all in Prussia from any of the insect calamities caused by moths between 1860–1880, whilst the pure woods of Pine suffered severely.

The beneficial effect of mixed crops in obviating extensive injuries from insect enemies is perhaps almost the only point concerning which German sylviculturists are absolutely unanimous. That such calamities as have from time to time occurred in different parts of Germany have not taken place similarly, though on a smaller scale, in any part of Britain is, most probably, solely due to the happy chance or the prudent sagacity of woodland proprietors imitating nature by the formation of mixed woods rather than pure forest over areas of considerable extent. Experience on a large scale in Germany has shown that the number of dangerous insects is not only less in mixed than in pure crops, but also that at the same time the number of insectivorous birds is greater. Should Sir Herbert Maxwell's advice be extensively adopted, however, the days of such comparative immunity will almost to a certainty be numbered.

What is true about dangers from injurious insects, may also be repeated as to windfall and breakage from storms, and to damage wrought by accumulations of snow or ice on the

[1] *Zeitschrift für Forst- und Jagdwesen*, 1881, p. 1.

crowns of trees. In these respects again, conifers suffer to a far greater extent than broad-leaved genera, among which Beech suffers more than Oak, Ash, Maples, or Elms. Of all forest trees Spruce is, owing to the shallowness of its root-system, most apt to become windfall; whilst Scots Pine is, owing to the brittleness or low degree of elasticity of its branches, more apt to have its crown injured by violent winds or heavy accumulations of snow or ice. A year seldom goes past without the necessity of registering damage from one or other of these causes; though, fortunately, destructive storms like the N.E. gale of November 17, 1893, in Scotland are rare.

To prevent, so far as possible, damage from storms, Hess[1] recommends the admixture of deep-rooting species along with those which have shallow root-systems, and adds that—

'The injuries caused by storms during the last decades should be a warning to those sylviculturists who are over-eager to transform forests of broad-leaved trees into coniferous woods; such transformations should only take place when justified by extreme necessity,'

as, for instance, when soils have become greatly deteriorated by faulty treatment.

Accumulations of snow on the crowns of trees become diminished in mixed crops, especially when the conifers are interspersed with broad-leaved trees. Besides allowing the snow to reach the ground, the bare crowns of the latter afford some slight support to the snow-laden crowns of the former. Even when breakage results, it is then more of an individual nature, and does not extend throughout large patches, or over the whole of the crop as sometimes happens in Spruce and Pine woods. Here again Hess lays down[2], as the first sylvicultural measure to be taken for the prevention of damage from snow and ice :—

'The avoidance of the formation of pure forests of species liable to suffer. Scots Pine is entirely out of place in such localities. So far as

----

[1] *Op. cit.*, vol. ii., 1890, p. 287.
[2] *Op. cit.*, vol. ii. p. 315.

possible, broad-leaved species of trees should be scattered, in sufficient quantity and at convenient distances, throughout Spruce woods.'

That similar drawbacks are also incidental to the formation of too extensive pure crops of either cereals or roots in agricultural utilization of the soil, may be seen by a reference to Fream's *Agriculture* [1], from which the following short extract may here be given :—

' In other words, side by side with the excessive, or exclusive, cultivation of one kind of plant, the pests—whether insects or fungi—which prey upon that plant may be expected to become more abundant, for they find their victims literally crowded together, and therefore extremely accessible.'

7. *Natural Regeneration of Mixed Woods is on the whole easier than in the case of Pure Crops.* This is but of minor importance compared with many of the other more substantial advantages offered by mixed crops. It is due to the special treatment and tending which must be bestowed on the crops in order that, so far as possible, each species may be placed in the natural conditions most favourable to its growth and further development. With some attention given towards securing for each kind of tree the soil and situation best adapted for it, a healthier growth, and a richer production of seed are natural results, owing to larger secretions of starchy reserves of surplus nourishment. Hence regeneration from seed in small groups and patches is comparatively easy, and is more likely to be successful than when one single species predominates over the whole woodland area notwithstanding the usually frequent changes with regard to the depth of soil and the quantity of soil-moisture. The species of tree which gains most in this way is again the Spruce, which cannot usually be regenerated under parent standards owing to danger of windfall; but, when treated thus in mixed crops, it can be reproduced without the danger otherwise threatened to young seedling crops by weevils (*Curculionidae*). The natural regeneration of the other great shade-

---

[1] *Elements of Agriculture*, 4th edit. 1892, p. 333.

bearing species, Beech, Hornbeam, and Silver Fir, offer in general but slight difficulty. In mixed crops these trees would usually be reproduced naturally and cleared away as soon as possible, thus leaving the young seedling crop to grow up as underwood under the light-demanding standards of other kinds of trees left to thicken into stems of fine marketable dimensions.

In the regeneration of mixed woods it will be almost invariably found advisable to plant out large transplants of the more valuable species of timber trees; for their disposal throughout the soil-protecting, ruling species, and the advantage in age and development which it is desired to give them over it, are much better ensured by this than by any other means. But, at the same time, wherever special conditions of soil render the natural regeneration of the light-demanding species desirable, this can also be effected; and any blanks that remain afterwards can easily be filled up with one or other of the shade-bearing and soil-protecting kinds of trees.

8. *By Admixture of Various Kinds of trees in one series of Crops, the operations of tending, harvesting, &c., can be conducted most economically.* This is another advantage of an almost axiomatic nature. Suppose, for example, that any landowner has 800 acres of woods, which he wishes to have stocked with the four kinds of trees, Larch, Scots Pine, Spruce, and Silver Fir, worked with a rotation of say 80 years each. If formed into pure forests, when once the crops were in full working order so as to yield annual falls of timber, there would need to be four distinct series of annual crops from 0–80 years old, in place of one series of mixed crops of the same ages; and all operations of thinning, tending, felling, extracting, and reproducing would have annually to be carried out in four different directions and over small areas of 2½ acres each, instead of being concentrated on one compartment or compact area of 10 acres in extent for each annual crop. This latter method represents a very great saving both in time and money; hence when

woodland estates are small, the formation of mixed crops is practically the only way of obtaining an annual outturn consisting of more than one kind of timber.

Such are the advantages which Mixed Woods possess over Pure Forests. And what can be urged against them? Nothing, except that their management needs a better knowledge of Forestry, and greater care and attention on the part of the forester.

In the above, Sir Herbert Maxwell's assertion that the formation of mixed woods is '*in itself a hindrance to profitable management*' has been disproved; and it has been shown that, on the contrary, they have many advantages over pure forests. By no means the least of these is the greater security with which the crops may be expected to attain their normal maturity. But, in sylviculture, there are no hard and fast rules applicable to all soils and situations; and circumstances can often occur not only to justify, but even to recommend, the formation of pure forests. The case has been very well put by Heyer in the following words [1] :—

' Notwithstanding the manifold and decided advantages which mixed forests offer in general, the formation of pure crops, to be permanently worked as such, is by no means precluded; for they in many cases offer such peculiar advantages as render it advisable or even necessary to retain or form them. Where, for instance, the local market or the nature of the soil and situation (e. g. when the soil is either very wet or very dry) is favourable to one particular species of tree only, that individual kind should be cultivated. But, at the same time, such cases are rather exceptional; *hence, as a rule, the formation of mixed woods is certainly to be recommended.*'

[1] *Der Waldbau*, 4th edit. 1891, p. 39.

CHAPTER VII

CONCERNING THE FORMATION AND GENERAL TREATMENT OF MIXED WOODS

In the preceding chapter it was shown that although in pure forests, where only one species of tree has mainly to be taken into account, each block of woodland may have the advantage of being easier to work as regards its tending, the harvesting of its mature crop, and its regeneration (though not necessarily, so far as the latter is concerned), yet such woods are much more exposed to climatic dangers like extensive windfall over areas covered with Spruce, or damage from snow and ice in Pine tracts, or devastation from noxious insects, as in the case of both Spruce and Pine, or fungoid diseases, as in Larch, Silver Fir, Spruce, and Pine. Pure forests cripple and confine the management to one kind of timber practically in any one locality; and if the market for that particular kind of wood falls so as to become unremunerative, it is not easy all at once to transform the woods into more profitable crops.

It will in general, therefore, be considerably to the advantage of landowners to avoid the formation of pure forests over extensive areas, though it may sometimes happen that there is very little free choice about the matter. Thus, when the soil is too poor for mixed crops of Oak, Ash, Maple, Sycamore, Elm, Larch, &c., to be grown along with Beech, Hornbeam, Spruce, and Silver or Douglas Firs, the choice is often confined to conifers alone, or on the poorest classes of soil to the Scots and Black Pines alone, if timber crops are to be grown at all; or undrained tracts may be better suited for Alders, Willows, or Poplars than for any other genera; or again, the circumstances of local marts may be such as to ensure a constant demand for one particular class of timber, which thus becomes the most remunerative crop under these special conditions.

In most mixed woods of about equal age certain species, either through greater energy of growth in height, or in consequence of soil and situation specially favouring their development, or from some combination of both these factors, attain such advantages in rate and extent of development as to induce or necessitate a corresponding decrease below the normal development in the other species. Hence sooner or later these latter trees, unless specially tended, have to be removed from the crop even although they may not have attained their physical and technical maturity. It thus happens that, in many cases, towards the end of the period fixed by the Working Plan for mixed woods, the mature crop consists mainly of one species; and this is usually that which has shown itself most capable of bearing shade. And when this species is being reproduced and harvested, advantageous opportunities are once more given for the formation of mixed crops by the planting out of stout transplants of the more valuable kinds of trees on whatever patches of soil seem most suitable to their healthy growth and vigorous development.

With their denser crops, their better conservation of the productive capacity of the soil, their higher percentage of stems of large dimensions, clean growth, and good technical properties, their power of satisfying demands for various kinds and qualities of timber, and their comparative security from calamities and devastations arising from either organic or inorganic causes, it can hardly be denied that mixed woods are more profitable, and rest on a much more secure financial basis than pure forests,—provided always that they are formed of the species for which the soil and situation are best suited, and that the timber grown is such as suits the requirements of the ordinary markets.   One of the highest authorities on the subject gives the following practical advice [1] :—

'Considering the circumstances of the timber market in general nowadays, and of local demands (i. e. in Germany), and considering also their incontestibly thriving growth, no argument is required to show that conifers, and in particular Spruce, must nearly everywhere claim the lion's share in the composition of the mixed forests of the future.'

For Britain, however, the Douglas Fir and the Menzies or Sitka Spruce hold out very much better financial promise than the common or Norway Spruce.

In the primeval forests of Britain mixed woods of Oak and Beech covered most of the low-lying tracts and the better qualities of soil on the uplands ; Beech was the principal tree on the limy hills of central and southern England ; whilst Scots Pine probably asserted itself over the poorer hillsides, with patches of Birch here and there.   On the richer, deeper soil Oak was of quicker growth than Beech, and, growing in localities eminently suited to its requirements, could well regenerate itself naturally under the shade of the parent trees, as may even now be seen in the case of pure woods formed on rich alluvial tracts.   The great forests of the Highlands of Scotland were composed for the most part of Scots Pine, because it was content with indifferent classes of soil where few other trees were able to dispute its foothold.   A thick growth of moss helped to maintain the productive capacity of the soil when the Pine began to thin itself about the twentieth to thirtieth year, and then became somewhat broken in canopy.   In our damp insular climate this not only diminished the danger of soil-deterioration through sun and wind, but it also favoured good natural reproduction and the formation of a closer canopy.   Scots Elm, Ash, Alder, Aspen, Willows, and Mountain Ash were all practically confined either by their demands for moisture, or for soil-fertility, or for both,

---

[1] Gayer, *Zeitschrift für Forst- und Jagdwesen*, 1892, p. 386.

to the ravines on the hillsides and the lower-lying tracts where the deeper and richer soil lay, and where the abundance of soil-moisture rendered a loose, broken canopy and a light annual fall of leaves matters of no particular moment—where often, in fact, superabundance of soil-moisture made the evaporating action of sun and wind beneficial rather than prejudicial. The wet soil, the severe winters, and the frequent late and early frosts in such humid localities safeguarded their domain from being encroached on by the Oak or the Beech. That most light-demanding, but hardiest, and most graceful, of all our indigenous trees, the Birch, was forced to seek its permanent home wherever no other genera of trees could assert a foothold and maintain themselves in growth.

Rules for the Formation of Pure Forests.

Whenever it is deemed advisable to form pure forests artificially, the following rules laid down by Heyer[1] should be noted as having proved themselves sound both in principle and in practice :—

I. *As a rule, only those species should be grown in pure forests, which retain or increase the quality or productive capacity of the soil. Such are :—*

1. *Those which are thickly foliaged, and grow in close canopy.*—These qualities are pre-eminently exhibited by the Beech, Spruces, Douglas Fir, and Silver Firs, and in a less degree by the Hornbeam, Lime, and Chestnut. Whether or not the Sycamore and the Maple may also be included in this class for Britain, depends to a great extent on the mineral strength and freshness of the soil in each concrete case.

2. *Those lightly-foliaged conifers which are evergreen.*—The thick, spongy growth of *Hypnum* mosses that springs up when such woods begin the natural process of thinning themselves,

---

[1] *Der Waldbau*, 4th edit. 1891, pp. 30–32.

and which acts like humus or mould in protecting the soil-moisture against the effects of sun and wind, gradually disappears when the canopy commences to become very much interrupted and broken. The utilization and reproduction of such pure forests should therefore not be delayed too long. The species coming under this description are the various members of the Pine genus, although Black (Austrian and Corsican), Weymouth, and Maritime Pines are all more densely foliaged, and maintain better canopy, than our indigenous Scots Pine.

II. *Such species as do not retain the productive capacity of the soil may, however, be grown in pure forests :—*

1. *When worked with low rotation.*—All species of forest trees protect the soil during the earlier stages of their development, because the crop is denser and the crowns of foliage are nearer the ground. This applies, along with other crops, to osier-holts and coppice-woods, for example.

2. *When the productive capacity of the soil is not endangered by imperfect cover.*—Under this head are classifiable such cases as marshy soil, where evaporation of the moisture by insolation and by the free play of winds is directly beneficial; or as valleys or low-lying tracts whose soil always remains fresh or damp, and whose depth and porosity would only be injuriously interfered with by an accumulation of humus. Alder and Birch are not infrequently to be found on soils of such description, and Ash, Elm, and Oak on those of a better class formed by rich alluvial deposits.

It may be noted that in the above no provision seems made for such trees as Larch and Oak in general. Pure forests of these species may of course be formed under II. 1; but, if they are to be maintained as pure forest to be worked at the ordinary periods of rotation for such classes of timber, they should certainly be underplanted with a shade-bearing species to protect the soil. Many of the Larch forests of Scotland would (from a sylvicultural point of view—but pasturage would then be out of the question) be much healthier and more vigorous in growth if they had been well thinned and under-planted with Spruces or Silver Firs about their twenty-fifth to thirtieth year. This is apparently anomalous, for Larch forests occur pure in its Alpine home, though only at high elevations, where the atmosphere is humid, where there is a strong tendency to the growth of moss even in such open forests, where the general growth of Larch is much more vigorous, where there is nearer approach to the maintenance of good canopy, and where, finally, the soil is protected by a deep covering of snow for most of the six to seven months of the year during which the tree is bereft of its foliage.

Rules for the Formation of Mixed Forests.

In the same way, general rules for the formation of mixed crops can also be framed in accordance with the laws of nature and the principles of sylviculture. Those tabulated briefly by Gayer [1] are as follows :—

1. *The concrete conditions of soil and situation must be such as are favourable to the normal development of all species of trees intended to be grown in admixture.*—It would be erroneous to suppose that mixed crops are advisable only on the better classes of soil, although of course the quality of the latter is an important factor in determining the choice of species to be admixed. In fact, it is on some of the poorer kinds of land that the greatest general advantages of mixed woods make themselves most apparent, particularly when the soils are of a sandy nature.

2. *The admixture of species must not be such as ultimately to endanger the productive capacity of the soil.*—This is practically a repetition of the first principle laid down with regard to the

---

[1] *Der Waldbau*, 3rd edit. 1889, pp. 216, 217.

formation of pure forests, and amounts, in other words, to the statement that the capital value of the woodland soil, i. e. its capacity for yielding interest or returns in timber under proper treatment, must not be tampered with, risked, or diminished by injudicious management.

3. *Throughout the whole period of rotation each species must find the due amount of growing-space and of exposure to light, air, and warmth suited to its requirements.*—This condition is not at all confined to the development of the crown, but applies equally to the extension of the root-system ; and it becomes of all the more importance the nearer the individual species of trees approach to their physical maturity.

Proceeding from the general considerations that the possibility of forming mixed crops of various species is dependent—

    i. On their individual capability of protecting or improving the productive capacity of the soil,

    ii. On their relative demands with regard to light, and

    iii. On their relative rate of growth in height,—

the following rules for the formation of mixed forests were formulated by Heyer[1] :—

1. *The principal or ruling species must be capable of improving the soil.*—That is to say, the matrix should consist of genera and species like Beech, Hornbeam, Spruces, Silver Firs, Douglas Fir, and Scots or Black Pines.

2. *Shade-bearing species may be grown in admixture with each other when their rate of growth is about equal, or when the slower-growing species is protected against the quicker-growing.*—The protective measure adoptable in the latter case consists of giving some start at first to the slower-growing species, of allowing it to form the great majority of the crop, or of favouring it at the time of natural reproduction, and later on by

---

[1] *Op. cit.,* pp. 45–57.

lopping, topping, or even cutting out the other species of trees threatening to suppress it.

3. *Shade-bearing (thickly-foliaged) species can be intermixed with light-demanding (thinly-foliaged) species, when the latter are of more rapid growth in height, or when they are afforded some advantage as to age or height.*—If, however, the shade-bearing species is to be expected to develop vigorously, instead of being suppressed during the youthful period of growth, it must form the principal or ruling species numerically.

4. *Light-demanding species should not be permanently associated together as timber crops.*—When the slower-growing species is at the same time relatively the more capable of bearing shade (e. g. Austrian Pine and Larch, Austrian Pine and Scots Pine), exception to this rule can be made :—

(*a*) On very good soil not exposed to deterioration under the light canopy of the thinly-foliaged species. Under such conditions mixed growth of Alders and Ash, Oak and Elm, Oak and Maples, &c., are not at all injudicious.

(*b*) On very poor (sandy) soil, principally given up to conifers, where of broad-leaved trees the Birch alone is content to grow.

Except on soils and situations where the Scots Pine shows a decided tendency towards a fuller degree of shade-bearing capacity than usual, a mixture of this species and Larch is not at all advisable, although very often made in Britain (see p. 155).

5. *The subordinate species should be introduced individually and not in patches or groups.*—Here, again, however, exceptions may be made :—

(*a*) When the quality of the soil is variable, so that patches here and there are specially suited for any one particular class of tree (e. g. Spruce on shallow, stony soil; Ash and Alder on damp, wet spots; Pines on dry patches).

(*b*) When a light-demanding species is to be grown along with a shade-bearing species of quicker growth (e. g. Oaks grown in admixture with Beech on soils specially favourable to the latter).

(*c*) When standards (e. g. of Oak) are retained for a second period of rotation. Here some protection is afforded to the soil until the young seedling crop forms canopy, and at the same time the formation of shoots from dormant buds along the stem, leading too often to '*stag-headedness*,' is prevented.

A.—Concerning Mixed Crops of Shade-bearing Species.

Spruce and Silver Fir.—Where Spruce forms the ruling species, the introduction of Silver Fir as part of the crop is of great advantage in diminishing the danger of windfall and of destructive attacks from injurious insects ; but where the soil is good enough for the Silver Fir to form the bulk of the mature crop, any admixture of Spruce has just the opposite effects. As in early youth Spruce is the more rapid in growth, the Silver Fir should be introduced in considerably larger numbers than is ultimately desired. Thus, if originally planted in the proportion of two of Silver Fir to one of Spruce, the mature crop would ultimately consist of about equal numbers of each kind of tree.

Beech and Spruce.—Although Beech is of somewhat quicker growth at first, it soon gets caught up by the Spruce, which then usually remains predominant. Unless the Beech is to be ultimately suppressed, the Spruce must only be scattered singly throughout the crop ; and even then some lopping of the branches will have to take place. Where stony outcrops occur in Beech woods, it is best to form pure clumps of Spruce.

Beech and Silver Fir.—At first the Beech develops more rapidly, and interferes with the growth of the Silver Fir ; but later on this overtakes and threatens to suppress the former.

B.—Concerning Mixed Crops of Shade-bearing and Light-demanding Species.

Common Spruce as ruling Species or Matrix.

During the youthful period of development all light-demanding kinds of trees are of quicker growth than the Spruce, and damage the latter by injuring the tender leading-shoots when it begins to catch them up in growth. Later on, most of the light-demanding species are overtaken by the Spruce; and unless it is intended to utilize them by removal during the periodical thinnings, the former require special tending and protection against the latter.

Spruce with Oak, Ash, Maple, Sycamore, or Elm as subordinate Species.—All these species are overtaken in growth by Spruce between the fifteenth to twenty-fifth year as a rule, although on very good soils they retain their advantage somewhat longer, especially when they also predominate numerically. For the most part, however, these valuable species should be grown in admixture with the Beech. Even when Oak has an advantage in growth up to about twenty years, it is still usually liable to be caught up when the Spruce gets to about fifty years of age; and then it is either suppressed, or else hindered in afterwards attaining its normal girth as mature timber. Experience shows that a mixture of Oak with Spruce should only take place when once the former has advanced considerably towards maturity. This can best be effected by cultivating the Oak in pure forest first of all, and then, about the sixtieth to seventieth year, making a strong thinning or partial clearance, with simultaneous sowing or planting of the Spruce. The whole crop is then harvested as one mature crop about eighty to a hundred years later on.

Spruce and Birch.—Of all the subordinate species of trees which are at first of quicker growth than the Spruce, the Birch does most damage, owing to the extent to which the whip-like twigs scour and injure the leading-shoots. Thus, although useful as a nurse in localities exposed to frosts, an admixture of Birch is not much favoured. The reason is more especially owing to its strong reproductive capacity from the stool; for it is extremely difficult to get rid of the latter tree when it is no longer desired as part of the crop.

Aspen and Willow are also prejudicial to the development of the Spruce, but not to the same extent as the Birch.

Spruce with Scots Pine.—During the earliest period of development the quicker-growing Pine affords the Spruce beneficial protection against frost and heat; but the Pine cannot exceed one-fifth to one-seventh of the crop without threatening to interfere for a long time with the normal growth of the Spruce. Later on the Pine, owing to its tendency towards a spreading crown, does more or less damage by rubbing and scouring the shoots of the Spruce. On some situations the Pine lags behind the Spruce in growth between the thirty-fifth to fiftieth year; but on good soil it regains the lost advantage, and continues till maturity among the dominating or pre-dominating classes. In other localities observations have shown that on somewhat dry sandy soils the Pine is of quicker growth throughout its whole period of development; whilst on fresh soils the Spruce shoots continuously ahead when once it has caught the other up. A permanent mixture of these trees is as a rule only recommendable on medium classes of soil; but by this means the Pine can be made to hold out longer periods of rotation than otherwise, and can develop into large-girthed stems of considerable marketable value.

Spruce with Larch—This mixture is at first very similar to that of Spruce and Pine, except that, on soils suited to it, the Larch remains permanently predominant. Where, however, the soil is wanting in depth or in mineral strength, and very frequently in low-lying localities, the Larch is apt to be over-taken in growth in height by the Spruce, and then it is hardly profitable to endeavour to retain the former so as ultimately to form part of the mature crop; it is better to cut it out at once.

Silver Fir as a ruling species closely resembles Spruce. But it is less active in suppressing the subordinate species when once it catches them up in growth, as its shade is somewhat less dense. The same will also most probably prove correct with regard to the **Douglas Fir.**

Black Pines, i. e. **Austrian and Corsican,** can bear a considerable amount of shade, and certainly improve the soil in no slight degree through their rich fall of needles; hence they are naturally adapted as ruling species, more especially when grown along with very lightly-foliaged trees like Larch.

Beech as ruling Species.

Writing in 1791, Gilpin spoke thus slightingly of the Beech :—

'The Oak, the Ash, and the Elm, are commonly dignified in our English woods, as a distinct class, by the title of timber trees . . . After timber trees, the Beech deserves our notice. Some indeed rank the Beech among timber trees; but, I believe, in general it does not find that respect, as its wood is of a soft spongy nature; sappy and alluring to the worm.'

But in what different estimation is this noble woodland tree held by the continental sylviculturists who have studied the subject most thoroughly, and have the best right to speak authoritatively concerning it !

Thus Prof. Gayer of Munich says [1] :—

' There are many localities in which Beech will continue to be a valuable wood from a financial point of view (i. e. for fuel in Germany). But where such may not be the case, it will still retain its insurpassable sylvicultural value; for *without the Beech there can no more be properly tended forests of broad-leaved genera,* as along with it would have to be given up many other valuable timber trees, whose production is only possible with the aid of Beech.'

[1] *Op. cit.* 1889, p. 448.

And Ney's opinion is as follows [1]:—

'The Beech, which in pure crops belongs to the more valuable species of trees only on account of its protective influence on the productive capacity of the woodlands, is of extraordinary value in mixed crops owing to its capacity for casting a dense shade over the ground and for improving the soil by its strong fall of leaves. When an admixture of Beech is judiciously formed, all species of timber trees show a much greater energy of growth than in pure crops. It safeguards the productive capacity of the soil better than any other species of tree, and therefore well deserves the name of "*the Mother of the Forest*" which has been given to it.'

Beech with Oak.—In exposed localities, or on shallow, and especially on limy soils, Oak is slower in growth than Beech, although on most other soils and situations it is decidedly of more rapid development. But, even in this latter case, the advantage won is seldom so great as to obviate the necessity for adopting measures to protect it against the Beech. This is in some cases effected by forming pure crops of Oak first of all, and then underplanting them with Beech either during the pole-forest stage of growth, or about the seventieth to ninetieth year, when the main growth in height has been completed. A heavy thinning or partial clearance previously takes place in each case. Again, in mature mixed crops, the Oak can be favoured by either being reproduced first in groups of 20 to 30 feet in diameter, or else by planting up the blank patches in Beech seedling-growth with strong Oak transplants. And, by favouring the Oak during all operations of thinning, or by stimulating it to increased growth in height through removal of the lower branches of the crown periodically, before they attain a diameter of over four inches, much can be done to assist its development.

On very good soil, not apt to deteriorate under the loose canopy of the Oak, this genus can be introduced to such an extent as to far outnumber the Beech ; but, on the less fertile

[1] *Lehre vom Waldbau*, 1885, p. 93.

and the merely average classes of forest soil, considerations regarding the conservation of the productive capacity of the land necessitate each Oak being surrounded with Beech. Where this is fully acted up to, there will be seldom more than eight to ten Oaks per acre attaining maturity at 100 to 120 years of age, though on the best classes of soil the number of Oaks in the crop may perhaps amount to as many as thirty to forty per acre. When the Oaks have been introduced in too large numbers at first, an early commencement should be made to reduce them gradually to the proper proportion they are intended ultimately to form in the mature crop. This can best be done by the removal of all stems of indifferent growth. If, in order to obtain very large-girthed timber, it seems desirable to allow the Oaks to remain on good soil for a second period of rotation of the Beech, they can be left in the numbers above given; but, before the end of that time, they will, in the latter case, have formed canopy, under which the Beech may still grow up as a tree, though not so as to attain normal dimensions. On average classes of soil eight to ten Oaks per acre will, at the end of the second period of rotation, overshadow about one-fourth of the soil, and sixteen to twenty about one-half of the area.

Beech with Ash, Maple, Sycamore, and Elm.—Despite a very considerable density of foliage in the Sycamore under certain circumstances, all these species strongly resemble the Oak in its attitude towards the Beech; for all are light-demanding and of quicker growth at first than the latter. This advantage they usually retain on deep, fresh soils; but otherwise they are, as a rule, caught up in growth by the Beech during the pole-forest stage of development. In many situations they are apt to sow themselves so freely as materially to interfere with the natural reproduction of the ruling, soil-protecting species; hence it is advisable to utilize them at the time of preparing the soil for the reception of the Beech-mast, and merely to introduce them here and there singly or in small patches throughout the new seedling crop.   By the use of healthy transplants, and by selection of the most suitable patches for the growth of these various species, better results can be attained than by endeavouring to reproduce the whole crop as mixed forest.   In addition to some advantage in age and height at first, careful tending of these more valuable species is requisite during all the operations of thinning.   It is only on the soils most favourable to their growth that these kinds of trees can hold out the usual period of rotation (90 to 120 years) best suited to the Beech; hence, by the time fixed for the fall of the mature crop, they will in many cases have disappeared from the woods, having been removed for disposal when they had reached their full physical and economic maturity.   The more the soil is suited for the Beech, the earlier will the fall of these subordinate species be necessitated; but, in any case, the quantity and quality of the timber harvested will yield better returns than if attempts be made to grow these kinds of trees in pure forest without the beneficial assistance of the Beech.

For moister classes of soil the **Hornbeam** can often yield better service than the Beech; and, as its growth is not so energetic, there is less likelihood of its catching up the nobler species and threatening to suppress them.   Although it gives much smaller returns in timber, the latter is, owing to its extreme toughness, valuable for machinery work.   Again, as it attains its physical and economic maturity much earlier than the Beech, it is in some respects even a more favourable ruling species than the latter,—for example, when it is desirable to allow the soil to remain as short a time as possible in the sole possession of the ruling species : for, practically, the natural reproduction of the Hornbeam could take place immediately after the fall of the subordinate species.

Beech with Birch, Aspen, and Willow.—As a rule no artificial efforts are necessary to attain an admixture of the soft-woods in seedling crops of Beech.   In fact they sometimes occur self-sown to such an extent as to come under the heading of '*forest weeds*,' for the removal of which a considerable expenditure of time, trouble, and outlay is often necessitated. Where they occur singly and individually here and there, not only is no damage done by them, but they may prove beneficial in protecting the slow-growing Beech against late and early frosts ; and, with favourable opportunity of disposing of small material, they can often be made to yield fair returns when cut out during the cultural operations of weeding and clearing. Wherever they occur in considerable numbers, however, they interfere considerably with the development of the Beech, and must be cut out as soon as possible in order to avoid the formation of blanks. On account of their strong reproductive capacity, and of the rapidity in growth of their coppice-shoots and stoles, it is better merely to lop them, and to leave snags of two or three feet standing, than to cut them flush with the ground. With the timely removal of such softwood nurses, a good opportunity is offered for the introduction of sturdy transplants of the nobler species of our forest trees. But when once the proper time for their clearance has been allowed to slip by, it is difficult to adopt any other measure than their very gradual removal during the subsequent operations of tending, without at the same time exposing the principal crop to the danger of being bent down or laid; for the young Beech poles are apt to be drawn up so fast as to become top-heavy without the partial support of the softwoods.

Beech with Scots Pine.—Each of these trees accommodates itself well to the peculiar requirements of the other, for the light-demanding Pine is throughout all the stages of its development of more rapid growth than the Beech. If introduced to a moderate extent only, Pine does little damage to the latter by overshadowing; whilst, at the same time, it acts beneficially as a nurse in protecting the Beech from extremes of heat and cold. In such mixed crops the Scots Pine attains more valuable dimensions, and is of better quality, than when grown either pure or in admixture with other conifers. And it can also be then worked with a higher period of rotation, without incurring any financial loss.

It is a curious fact that, under the light shade of the Pine, the natural reproduction of the Beech can be more easily and successfully carried out than under the shelter of parent standards themselves.

As the natural habit of the Scots Pine tends towards an early lateral development of the crown, it should not be given any advantage in age or height to begin with ; its introduction can therefore easily take place either by sowing or by planting here and there throughout the young seedling crops of the ruling species.

The **Larch**, as a subordinate species in admixture with Beech, on the whole closely resembles the Scots Pine, except that it then often reaches its physical and economic maturity at about fifty to sixty years of age, and that it makes higher demands on the quality of the soil.

C.—Concerning Mixed Crops of Light-demanding Species.

Considerations relative to the conservation of the productive capacity of the soil demand that under ordinary circumstances some shade-bearing species should form a portion of the crop; for the true scientific principle is, as has already been stated above, that it should, in fact, form the major part of the crop, i.e. it should be the ruling species. But, under certain special conditions as to soil and situation, it may seem desirable to the proprietor to depart from this principle, and to form mixed forests consisting entirely of light-demanding trees. Such crops must ultimately, however, from their natural habits of growth, fail to maintain close canopy; hence, unless the soil is possessed of such physical properties as render it little likely to deteriorate through insolation and the action of winds, the practical formation of a mixed crop, by means of underplanting with the shade-bearing species for which the soil is best suited, is rendered advisable. Where fresh or moist, low-lying tracts with good soil are still under forest growth, mixed woods of Oak with Ash and Elm, or with Alder and Birch,—Alder with Ash, Birch, and Aspen,—Pines with Aspen and Alder,—and the like, are often to be seen in luxuriant growth and yielding good returns to the proprietors.

In Britain, the light-demanding species which is most frequently to be found forming the ruling species or matrix is undoubtedly the Scots Pine. But, even in our damp insular climate, the admixture with it of some species capable of bearing shade has a very beneficial effect on the soil, and through that on the quantity of timber obtained at the fall of the crop. For the better classes of Pine soil the Beech or Silver Fir, or on moist localities exposed to frosts, the Hornbeam and Douglas Fir, and for average and inferior localities the Spruce, Black (Austrian and Corsican) and Weymouth Pines are recommendable as subordinate species ; but as, on the inferior tracts classifiable as Pine soil, these trees are less tolerant of shade than elsewhere, they must receive a fair amount of consideration during all operations of tending and thinning. The admixture of Larch with Scots Pine on such localities indicates want of knowledge of the natural requirements of the Larch with respect to fertility and depth of soil ; whilst at the same time, owing to the more rapid growth of the Larch, the Pine suffers even from the light shade cast over it by the former.

Recapitulation.

As this subject is of great practical importance it may be well to summarize briefly what has been said at length in this and the preceding chapter in exposing the fallacy contained in Sir Herbert Maxwell's statement quoted on page 115. Where only one species of tree has mainly to be taken into account, each block of forest may have the advantage of being easier to work with regard to tending, to the fall of the mature timber, and to the reproduction of the crop (though not necessarily, so far as the latter is concerned), but it is much more exposed to climatic dangers like extensive windfall over areas covered with Spruce, or to damage from snow and ice in Pine tracts, or to attacks of insect enemies as with both Spruce and Pine, or to cankerous fungoid diseases as in the case of Larch or Silver Fir, &c.

Pure forests cripple and confine the management to one kind of timber practically in any one locality; and if the market for that particular kind goes down and becomes unremunerative, it is not easy all at once to pave the way for the production of other and more profitable woodland crops.

It will in general, therefore, be to the advantage of the land-owner to avoid the formation of pure forests over extensive areas. But occasionally it happens that there is little practical choice about the matter. The land may be too poor for Oak, Ash, Maple, Sycamore, Elm, Larch, &c. to grow in mixture along with Beech, Hornbeam, Spruces, and Silver or Douglas Firs, and then Pine forests are the best crops that can be formed if the land is to be put under timber at all; or undrainable tracts may be better suited for Alders or Willows and Poplars than for other genera; or local requirements may be such that a fairly well-assured and constant future market for one particular class of timber seems to point to its cultivation as most advisable and attractive.

In most mixed crops of about equal age certain species, either through greater energy of growth in height, or in consequence of soil and situation specially favouring their development, or from a combination of both these factors, attain advantages in rate and extent of development which induce or necessitate a corresponding fall below the par of normal development in the other species; hence sooner or later these latter species, unless specially tended, have to be removed from the crop, even although they may not have attained their physical and technical maturity. It thus happens that in many cases mixed woods must, towards the end of the period fixed by the Working-Plan for the fall of the mature crop, consist principally of one species, which is usually that most capable of bearing shade. But when this species has been either naturally or artificially reproduced, the most advantageous opportunities are again offered for the re-formation of mixed crops by the introduction of stout transplants of the more valuable timber-producing trees of other genera or species.

With their denser crops, their better conservation of the productive capacity of the soil, their higher percentage of stems of large dimensions, clean growth, and good technical properties, their power of satisfying demands for various kinds and qualities of timber, and their comparative security against calamities and devastations arising from organic or inorganic causes, it can hardly be denied that mixed forests are more profitable, and rest on a much more secure financial basis, than pure forests,—provided always that they are formed of the species of trees for which the soil and situation are best suited, and that the timber grown is such as meets the requirements of the ordinary markets. In fact, as compared with pure forests, mixed woods exhibit the economic paradox of holding out fair promises of a higher rate of annual interest, combined at the same time with better security for that very considerable portion of the capital which is represented by the growing stock of timber throughout the woodland area.

CHAPTER VIII

CONCERNING THE FORMATION AND REGENERATION OF WOODLAND CROPS

'Where any system of management satisfies all reasonable expectations, there can be no object in changing it. . . . . Not infrequently,
however, in one and the same forest, certain operations are carried out
in a masterly manner, whilst others might be better performed; and
observations and experiences made in one locality often remain unknown
for a long time in another.'—Burckhardt's *Säen und Pflanzen*, 1854
(1st edition), Preface.

Choice of Species of Trees to form the Crop.

WHETHER viewed from a purely sylvicultural standpoint
relative to the production of the largest quantity of timber of
the best quality, or from the financial standpoint relative to
the obtaining of the largest annual returns from the capital
represented by soil and growing stock, the choice of the kinds
of forest trees to form the crop is of the first importance under
all circumstances. Not every given soil and situation will produce any particular kind or assortment of timber; hence the
relegation of the various genera of woodland trees to the tracts
most likely to afford them a suitable permanent home is one
of the first duties of the sylviculturist. When dealing with
indigenous trees, or with such as have become naturalized by
long association with our native sylva, nature never makes
mistakes; and where any self-sown genera show good growth,

it may in general be accepted that soil and situation are favourable for their cultivation there. But this general rule is of course capable of being followed only in certain cases, and need not be always acted on—as, for example, when a spontaneous growth of Birch or Aspen asserts itself on soil from which the owner is entitled to expect better returns, and a more valuable yield of timber, than in the ordinary nature of things these kinds of crops can possibly promise.

The selection of trees is, however, by no means a free one; for there are many soils and situations where it is decidedly limited, or on which it is practically a case of 'Hobson's choice.' Thus, on dry sandy soils, no forest tree can compete with the Scots Pine, whilst on wet, sour, marshy land the common Alder holds equal sway; on peat bogs, that have been drained, a mixture of Scots and Weymouth Pines and Birch often offers the most advantageous crop; on low-lying tracts, liable to inundation annually, Willows and Poplars are best endowed by nature for growing under these special circumstances; whilst Spruce and Larch thrive in elevated localities with cool, humid atmosphere, above the zone where deciduous broad-leaved trees could find their natural and congenial habitat.

On many soils and situations a considerable number of genera of forest trees find conditions fairly well suited to their general growth; but, in general, one or other of the physical properties determines the elimination of certain kinds when crops are to be formed for profitable timber-production in contradistinction to ornamental growth and aesthetic effect. Thus, deficiency of soil-moisture may frequently preclude the idea of forming crops of Alder, Ash, Maple, Sycamore, and Elm, whilst want of atmospheric humidity may militate against the full normal development of Beech, Hornbeam, Spruce, and even Silver Fir; or, where the soil is wanting in depth and penetrability, Oak, Ash, Maples, Elm, Lime, Larch, Pines, Silver and Douglas Firs cannot extend their root-systems so as fully to utilize its nutrient soluble salts. It also happens that the favourable admixture of a certain mineral element in the soil affords special advantages to particular classes of trees— as, for instance, the better growth of Beech, Maples, Ash, and Black Pines on soils that are of a limy composition. And, in a certain degree connected with the conditions as to atmospheric humidity and soil-moisture, danger from late frosts in spring, or from early frosts in the autumn, can be so great as to preclude the planting out of Oak, Beech, Ash, Silver Fir, and even Spruce without a certain admixture of quick-growing softwoods as protective growth and nurses. Again, where heavy falls of snow are apt to occur, the brittle Scots Pine is less able to support the masses of snow and ice accumulating on its loose crown than other more elastic evergreen conifers.

And when one comes to the practical question, not only as regards the suitability of soil and situation for any given genus, but also with respect to their suitability for one particular species above all others, or with a view to the selection of the best genera for the formation of mixed crops, which have invaluable advantages over pure crops (see Chapter VI), the difficulties connected with the choice become increased. Thus the cultivation of the pedunculate Oak, Larch, Spruce, and Silver Fir will only prove satisfactory on soils and situations in every respect suited to their normal requirements. But over every large woodland tract there are variations, in quality of soil and degree of soil-moisture, which must be carefully observed and taken advantage of in the formation of mixed crops, instead of attempting to carry out the mixture in anything like a rigid or stencilled manner. No better example of want of prudent consideration, or want of knowledge, of the natural requirements of any species of forest tree can be given than to point to the mistakes made in the latter half of last century and the beginning of the present with respect to the extensive introduction of the Larch into Scotland. These errors have only had their natural results in diseased growth of the trees and disappointed hopes as to the financial returns anticipated.

When soil and situation seem about equally favourable to more than one kind of tree, prudential reasons will point to the formation of mixed woods of those genera. And considerations relative to the conservation of the productive capacity of the soil will naturally lead to a shade-bearing, soil-protecting kind forming the ruling species or matrix ; whilst the admixture of subordinate genera may be formed either individually or in small groups or patches, to such an extent as promises to be most remunerative until the final clearance of the crop takes place on its ultimately attaining maturity. Of course, when certain woods and assortments of timber have a good local market, the ease and readiness with which the timber of particular kinds, or classes, or qualities may be disposed of, will naturally lead the proprietor to their cultivation, even in preference to other genera for which the physical qualities of the soil and the local peculiarities of the situation are perhaps in reality better adapted.

Owing to its comparatively low value as timber, the Beech finds its place as the ruling tree solely with a view to the conservation of the productive energy of the soil, and need not be cultivated to any greater extent than this consideration demands. But the indirect benefits thus conferred on the lightly-foliaged trees grown in admixture with it enhance considerably the returns obtainable in quantity, quality, and monetary value from crops of Oak, Ash, Maple and Sycamore, Elm, Pines, and Larch formed together with Beech.

Where financial considerations have more weight given to them than those concerning the outturn of timber of special value for certain technical purposes (e. g. Oak for ship-building before the days of Teak-lined iron steamships) there can be no doubt that, for high-forest crops, the conifers, and, in particular, mixed woods of Spruce, Douglas Fir (a species very strongly to be recommended), Menzies Spruce, Silver Fir,

Noble Fir and Pines, yield a much larger outturn in timber, and, despite less value per unit of bulk, a higher percentage on the capital involved than can generally be obtained by the growth of mixed crops of deciduous forest trees.

At the same time, on suitable soil, coppices of Alder, Oak, and Willow may show financial results far exceeding those obtainable from conifers; but, in such cases, the quality of the soil and its location are usually very different from those of the areas which have naturally been set apart for sylviculture, owing to their unsuitability for arable or pastural utilization.

Where all genera seem to have about equal chances of thriving, pure forests of Douglas Fir, Silver Firs, and Spruces offer the greatest financial advantages, owing to the large quantities of timber produced and the large percentage of it utilizable for the higher technical purposes; Oak, Ash, Elm, and Maples come next; then Pines, softwoods, and last of all the Beech. These, however, are merely rough generalizations; for in each concrete case local climatic conditions exert their influence. Hence, in warm southern districts, Oak and other light-demanding genera yield more favourable results than the densely-foliaged Spruce, which thrives best in the cool, humid atmosphere prevalent within mountain ranges and at high elevations.

In technical value and financial returns obtainable per unit of volume, none of our indigenous timbers equals the Oak; whilst, for furniture and ornamental purposes, Ash, Maples, and Red Elm approach nearest to it. For the manufacture of agricultural implements, and for similar purposes requiring toughness combined with lightness, Ash timber has a value specially its own. Among conifers, good, sound, large-hearted Larch is of highest value, whilst Pine, Spruce, and Silver Fir vary locally in price. But as Spruce and Silver Fir are so much more productive per acre than other conifers, the large supplies of their timber annually thrown into the continental market may account for the comparatively low prices. The monetary returns available in general from the softwoods depend to a very great extent on local conditions of the timber market; whilst the Beech usually yields comparatively poor returns, despite its being the best, and the most heat-producing kind of fuel obtainable from woodlands. Its timber is, however, now coming largely into demand again.

Whilst the advantages derivable from mixed forests—including conservation of the productive capacity of the soil by the shade-bearing, ruling species; better utilization of the soil through differences in root-system, and of the light, air, and warmth by the crowns being generally at slightly different levels; protection against insect enemies, fungoid diseases, and damage from inorganic causes, &c.—are only attainable to the fullest degree when the admixture of genera occurs in individual stems, yet the casual differences in the physical properties of the soil can only be fully availed of when admixture takes place in patches or groups varying in extent with the variations in the soil. The general principles guiding the choice of species in mixed crops, apart from the fundamental rule regarding conservation of the soil applicable alike in the formation of pure and of mixed woods, are, broadly stated, that the conditions of soil and situation must suit all the kinds of trees to be intermixed, and that the more or less normal development of each species should be interfered with as little as possible by the others. Hence mixtures of trees with similar requirements as to light, or nearly equal capacity for bearing shade, usually recommend themselves only when one kind can be utilized so much earlier than the other as to permit of underplanting (e. g. Oak, with Ash, Maples, and Elm on good soil), or when, besides producing a greater quantity of timber with a higher percentage of technical and consequently of monetary value, other advantages are gained in protection against violent storms or the ravages of insects (e. g. when Spruce is mixed with Beech and Silver Fir). Thus, on soils and situations suitable for the Silver and Douglas Firs, an admixture of Spruce could only endanger the stability of the crop and increase the risk from injurious insects; whilst, on soils naturally best suited for Spruce, the introduction of these other genera can be of considerable advantage in the protection afforded against windfall and insects. The formation of mixed forests of Scots Pine and Larch, so often hitherto favoured in Scotland, is not one that can be recommended; for it is not in conformity with the principles of sylviculture. And practical experience has shown that the theoretical point of view in this matter is correct.

The most advantageous results are obtainable when a happy choice has been made combining light-demanding and shade-bearing genera, and when the former retain for a long time the early advantage they gain owing to their more rapid growth during the earliest stages of development. Where it is suited to the soil, Beech, owing to the very superior quality of leaf-mould formed by its dead foliage, or Hornbeam, yields the best results as matrix from a purely sylvicultural point of view. But, in Britain, financial considerations undoubtedly deserve greater attention from the land-owning classes than the scientific sylvicultural principles that ought to underlie the treatment of the State lands and Crown forests if they were properly administered; hence Spruces, Silver Firs, Douglas Fir, and in certain cases Austrian and Corsican Pines, are the forest trees most deserving of cultivation. Even though these species may be to a certain, and often to a considerable, extent influenced in their development by the other species of more rapid growth, yet even then they are still capable of yielding timber of good marketable quality and dimensions. Where, however, the interference with their normal development is only comparatively slight, the quality of the timber of the above kinds of forest trees decidedly gains by admixture.

And when once the choice of species has been made, the proportion which they shall each bear to the total number of plants per acre requires equally careful consideration. The shade-bearing, soil-improving kinds of trees must of course at first form the ruling species numerically; but in course of time they may become so reduced in number as, towards the end of the period fixed for harvesting the ultimate crop, to be in a minority.   For example, in crops formed by Oak, Ash, and Maples, with perhaps a few softwoods here and there, throughout a matrix of Beech, the processes of clearing, thinning, and tending throughout the various stages of growth of the crop will in most cases have led to the elimination of the softwoods between the thirtieth to fiftieth years, and of the Ash, Maples, and other similar woods between the sixtieth to eightieth years; whilst, from that time onwards, the main crop will consist of Oaks, with just a sufficient proportion of Beech to maintain the soil against deterioration.   And if the health and the vital energy of the Oak trees seem to justify their retention as a remunerative crop for a prolonged period of rotation, the Beech can be naturally reproduced below them as an underwood, or artificial sowing or underplanting can be resorted to, in order to prevent the soil from deteriorating or diminishing in its productive capacity.

Choice of Form and Density of the Crop.

The choice of the most advantageous form of crop to be grown on any woodland soil is determined partly by the genera selected for cultivation, partly by the nature of the soil and situation, and partly by the requirements of the land-owner. Should the proprietor have only a limited area under wood and yet desire to obtain, if possible, an annual outturn, he will naturally favour such methods as yield quick returns and only involve the locking-up of least capital in the way of soil *plus* growing stock.   And where bark for tanning, or withes for basket-making, are at all well paid, on soils that are favourable to the growth of Oak and Willows these forms of crop are amongst the most remunerative that can be grown.   On low wet tracts, too, Alder-coppice worked with a rotation of twenty-five to thirty years gives very excellent results, as the demand for this light wood for cigar-boxes and similar requirements is constant, and is not likely to be replaced in the immediate future by any other kind of wood, or by any substitute of a different nature.

High-forest of trees grown from seed produces absolutely higher monetary returns than ordinary methods of coppice ; but, owing to the greatly increased capital locked up in the soil and the timber stock ranging from the yearling growth to the mature crop, the financial result is often not in favour of the former. Again, in consequence of the bulky nature of forest crops, and the heavy costs of transport, local market conditions have such immense influence in determining the remunerativeness of any particular crop, that no general dictum can be laid down on the matter. This much, however, can be asserted, that, where woodlands are of considerable extent, mixed coniferous crops, worked with a rotation varying from 70 to 100 years, are in general most remunerative where the soil is of the average, or below the average, in quality; whilst, on soils of the better class, where conifers have a predisposition to suffer from fungoid diseases in the roots and stem, mixed crops of Beech, with a suitable large admixture of the nobler species of deciduous trees, really offer the safest and the best investments in the long run.

With the endless combinations of the various factors determining the rate of growth of, and the quality of timber produced by, woodland crops, the formulation of anything like hard and fast rules must be out of the question. The system of *total clearance and artificial reproduction* is simplest and easiest to carry out. The young growth is not endangered by the gradual removal of any parent standard trees ; a freer hand is allowed for the intermixture of subordinate species ; and, where it pays to undertake it, the grubbing up of the large quantities of fuel represented by the root-systems is a very simple matter. But, on the other hand, there is much greater danger to the young crop from rank growth of weeds, and from insect enemies (*Melolontha* and *Curculionidae* especially) ; whilst it is also more difficult to recognize and to deal with the variations in the quality of the soil with a view to making the best choice of genera for mixed crops. Except on good land and with kinds of trees hardy against frost, this method of treatment of the crop should only be adopted, on soils below the average in quality, when special considerations regarding either the land or the kind of trees to be reproduced render inadvisable the process of *gradual clearance of the crop with simultaneous regeneration under parent standards*. Where the soil is shallow or is apt to be easily heated, and to become deteriorated, as is often the case when the percentage of lime contained is great, the first point must be to maintain the land against the risk of deterioration ; hence the form of gradual clearance will be advisable only to so small an extent as to consist merely in the *selection and utilization of mature trees, individually or in small patches, without materially interrupting the density of the canopy*. Similar treatment will also be necessary in all localities exposed to violent winds, or wherever the woodland covering is essential for the maintenance of a good permanent supply of moisture in the soil. Wherever sylviculture is practised on a considerable scale, coppice-woods are practically confined to mild situations, and to shallow soils unsuited for the normal development of the larger root-systems of trees when growing in high forest.

For the production of the more valuable assortments of timber, growth in high forest is a necessity in the case of conifers. It is also generally advisable with regard to broad-leaved genera, although *the composite form* or *copse*, in which light-demanding trees of various age-classes form the standards over a coppice-growth of shade-bearing genera, is often highly remunerative where any favourable market exists for the produce yielded by the latter.

The density of the crops is a factor which exerts no little influence on their development and ultimate remunerativeness. Crowded woods have practically no greater total enjoyment of light, warmth, atmospheric food, and soil-nutriment than less densely packed crops growing in closed canopy; and as the production of timber is distributed over a smaller number of individual stems, the share which each is capable of receiving is greater than when the total of the available food-supplies has to be divided by the larger number. At the same time, in close-canopied, but not crowded, woods, the natural selection of the predominating stems to form the future mature crop proceeds more rapidly, as the individuals of forward growth utilize somewhat above their average share of light, &c., and consequently have a larger annual increment, than if the struggle for life and domination were more prolonged by having to be waged against a larger number of individuals of equal vigour. Where, owing to wide planting, the canopy is not of full normal density, the productive capacity of the soil and of the atmosphere is not utilized economically; hence a loss in timber takes place, which might easily be avoided. The earlier the young growth forms close canopy, the thinner are the branches formed, and the sooner do they die and drop off. A normal density of canopy therefore increases the technical value of timber by the production of clean boles having a high form-factor and approximating, more than otherwise would be the case, to the cylindrical shape represented by 1·0. All crops intended for timber production should, therefore, be maintained in close canopy till they have entered the pole-forest stage of growth. This is more particularly important with coniferous trees, whose technical, and consequently monetary, value is dependent, to a considerable extent, on freedom from hard horny branches and knots. And it is of course all the more necessary in the case of shade-bearing genera, whose lower branches are longer retentive of life. Where, therefore, the early attainment of full normal canopy can be achieved without special outlay at the time of the young crop being formed, it is undoubtedly of great financial advantage to see that due steps are taken to secure this end. Now, with this object in view it is by no means necessary to crowd plantations; for it is quite sufficient if the twigs and branchlets die off before becoming strong enough to leave snags or rotten ends in the timber, and if they can be displaced by the new annual zones of wood. This period varies with the different kinds of forest trees. The branchlets of the ring-pored, broad-leaved species of trees rot much sooner than those of conifers; small twigs of Oak or Beech snap off through their own weight when they have been dry for two or three years; whilst similar twigs of Spruce remain often as snags for ten or fifteen years, and get partially embodied in the stem if not removed. For the production of clean boles, therefore, conifers, and in particular the shade-bearing kinds, should be maintained in close canopy considerably longer than broad-leaved trees.

When the normal density of canopy is exceeded—that is, when young woods are crowded—the development of the individual stem is injuriously affected; for the crowns are liable to be drawn up so rapidly as to exceed the normal proportion which in well-developed trees should exist between height and girth. Crowded thickets should consequently be weeded and thinned as diligently as possible, even though this may involve considerable outlay. Otherwise the maintenance and prolongation of the individual struggle for supremacy on the one hand, and existence on the other, may have a very prejudicial effect on the ultimate remunerativeness of the crop throughout all the later stages of its development.

Where available, normal yield-tables in so far furnish hints regarding the most advantageous density or initial number of plants per acre in pure woods, that they show what the minimum number should be at the various ages for the particular genus of tree forming the crop on a soil of similar average quality. Supposing, for example, it were found that in Spruce forests, growing on soils of medium quality, the normal and most advantageous degree of density of canopy is attained by about 3,000 stems per acre, then it stands to reason that the formation of young crops, by planting at a distance of 4 ft. by 4 ft. (2,722 per acre), would entail loss by delaying the formation of canopy; whilst planting at 3 ft. by 3 ft. (4,840 per acre) might, even after allowing for casualties, go to the other disadvantageous extreme of crowding the plantation. Many of the older plantations in Scotland that I have seen were undoubtedly carried out without due consideration being given to the desirability of having canopy formed within fifteen to twenty-five years after the formation of the crop; hence the productive capacity of the soil cannot possibly have been utilized to its fullest extent.

A little consideration will show that even in the case of plantations which have been slow in attaining full canopy, i. e. which have not been formed of the most advantageous density, the crop may still form overcrowded woods between twenty to forty years of age, owing to the tendency to ramification induced by the individual plants having had larger growing-space than was requisite, or would have been dictated by sound economical considerations. Want of knowledge of the most advantageous density for young crops is undoubtedly one of the great faults of British sylviculture.

For planting, subject of course to variations according to the nature of the soil and the situation, Gayer[1] recommends the following distances :—

For shade-bearing genera and the Oak.

Class of Plants.	from	to
	ft.	ft.
Seedlings under 8 in. high	1 × 1	2 × 2
Small transplants, from 1 to 2 ft. high . . .	2 × 2	2¾ × 2¾
Stout transplants, from 3 to 4 ft. high	2¾ × 2¾	4 × 4
Very large transplants, from 6 to 8 ft. high . .	4 × 4	10 × 10

[1] *Waldbau,* 3rd edit. 1889, p. 355.

For light-demanding genera.

Class of Plants.	from	to
	ft.	ft.
Seedlings (Scots Pine, Larch) under 8 in. high .	$1\frac{3}{4} \times 1\frac{3}{4}$	$3\frac{1}{4} \times 3\frac{1}{4}$
Medium transplants, from 2 to 4 ft. high . . .	$3\frac{1}{2} \times 3\frac{1}{2}$	5×5
Stout transplants, 6 ft. high and above . . .	5×5	as desired

Of course, when close planting is carried out, the young plantations and thickets require very frequent and careful tending in the way of clearings, before the regular thinnings can be begun.

With reference to sowing, the amount of clean seed of average quality recommended by Ney as sufficient for the production of a moderate density of crop on soils of fair average quality, may be seen on reference to the author's *British Forest Trees*, 1893, page 50.

Choice of Method of Formation or of Reproduction of the Crop.

The great advantage derivable from artificial reproduction of forest crops consists in not being in any way dependent on the occurrence of favourable seed-years, as nowadays seed of all sorts can easily be obtained of good quality from nurserymen in whatever quantity may be desired for sowing out direct or for rearing seedlings and transplants in nurseries. That greater uniformity and regularity of the crop can be obtained artificially than when regeneration takes place naturally, either under parent standards or from seed shed from mature trees flanking the area to be re-wooded, hardly requires demonstration. As the methods of sowing or planting are usually simple, and as the operations are confined to a smaller area than when several annual falls are grouped together for gradual clearance, supervision and control are undoubtedly at the same time easier.

This seems a not unsuitable opportunity of noting one great difference between human beings and trees of the forest, and of pointing out that there is no such thing as hereditary disease in tree-growth.   Thus Professor Hartig says [1] :—

'The hereditary transference of diseases to succeeding generations is unknown in the vegetable world.   The seed of plants inflicted with all possible sorts of diseases may be utilized, without the slightest concern, for the formation of new crops.'

Where woodland crops are to be formed for the first time, a choice exists only between sowing and planting, in connexion with which—leaving out of sight the special requirements of the various genera and individual species of forest trees—certain general considerations require to be weighed.

With regard to the soil, experience has shown that, on places unfavourable to the early development of young crops, planting is preferable to sowing, owing to the greater sensitiveness and need of protection of young seedlings during the first stages of their existence.   And better results are usually obtainable from planting than from sowing, both on very damp, wet, cold, or stiff soils with a tendency to being lifted by frost, and on very loose soils apt to dry up easily, or such as may have become deteriorated superficially through insolation and exposure to exhausting winds, or which are liable to inundation, &c.   Where, owing to rank herbage of grass and weeds, young growth has to struggle for its very existence, sowing is the exception, and planting the rule, more especially when the genera of trees forming the young crop are of slow initial development.   Unfavourable situation with regard to climate, by retarding the growth during the first few years, also weights the balance in favour of planting, especially in raw, damp localities exposed to frost.   Where planting up is to take place on tracts that have been drained, but which are still damp enough to show a strong growth of rank weeds, sturdy transplants of hardy species, little sensitive to frost

---

[1] *Lehrbuch der Baumkrankheiten*, 2nd edit. 1889, p. 16.

(Pines, Birch, Aspen, Alder), should be planted out during autumn with balls of earth attached to their roots, or even tumped on mounds if the soil is actually moist. For, as Hartig well remarks [1]:—

'When trees or shrubs are planted out in any year and their natural process of development has been so much interfered with that the new shoots are not thoroughly developed by the time frost sets in, i. e. that the process of forming woody tissue has not been properly completed, such plants possess an abnormal disposition towards damage from frost. They may hold out mild winters ; but, if severe cold occurs, the plants may be killed outright.'

On all soils that are merely fresh and of a light, mild consistency—the happy mean between loose and stiff, neither apt to become too heated nor too rapidly cooled, and having no immoderate tendency towards rank growth of weeds—sowing is principally adopted, as also on rocky, stony outcrops where there is hardly sufficient soil for the proper carrying out of planting operations.

On the continent, sowing was formerly most generally practised, and it was not until the introduction, on an extensive scale, of the method of total clearance with artificial reproduction, that the present preference for planting became general abroad. In Scotland, the total destruction of the Pine woods, originally clothing vast extents of mountain sides now barren, naturally led to the artificial formation of forests wherever the proprietors desired to grow timber. And, in the vast majority of cases, the conditions of soil and situation—raw northern climate, rank growth of heather, heath, and other weeds, and deterioration of the surface-soil by long exposure to the effects of sun and wind—naturally pointed to planting as the best, and often the only, means of attaining the object in view. Good nursery seedlings, and especially sturdy transplants, must have fewer difficulties in establishing themselves than tender seedlings are exposed to in germinating on the area and over-

---

[1] *Op. cit.*, p. 15.

coming the disadvantages with which they have to struggle during the first two or three years of their existence.

Where sowing can take place under the shelter of standards, or in the lee of crops nearly mature, it is much more likely to be satisfactory than in the open. But, under nearly all circumstances, there is usually a good deal of work and outlay required for the filling up of blanks ; hence the final cost of the formation of such young crops is not always less than if planting had been carried out over the whole area at the very outset.

Until nursery seedlings or transplants have established themselves in their new abode, there is always a disturbance and a diminution of the activity of the root-system ; and this is accentuated by the trimming often requisite even in young plants, and always necessary in older transplants. In this respect sowing certainly has the indisputable advantage over planting of permitting a more natural and uninterrupted development of the root-system, and of effecting a better accommodation of the latter to the nutrient characteristics of the soil. The disturbances occasioned in transplants vary according to the species of tree, the method of planting, and the nature of the soil. Shallow-rooting species with good reproductive capacity establish themselves much more rapidly than deep-rooting species; while, on fertile soils, the efforts made towards accommodating themselves to the new conditions are more quickly responded to than on soils of merely average or indifferent quality.

By the use of transplants, too, many of the dangers from insect enemies during the first years of growth are avoided, although experience in many different localities has shown that sowings suffer on the whole less from *Curculionidae* than where plantations are formed with small seedlings.

Comparisons between crops formed by sowing and by planting during the last half-century have shown[1] that in

---

[1] Gayer's *Waldbau*, 3rd edit. 1889, pp. 383, 384.

very many cases plantations yield the better results both with regard to height and to girth ; whilst, as regards total production of timber (inclusive of thinnings), there is no particular difference, although the proportion of branches is larger in plantations owing to the greater initial growing-space enjoyed by each individual stem. Whether or not the advantages as regards dimensions will be maintained by plantations up to the time of their fall as mature crops, is a question for the answering of which no data are yet available for a trustworthy comparison.

On the following point, however, it is well to note the words of so eminent an authority on sylviculture as Professor Gayer of Munich :—

'That the rapid initial development of many plantations considerably affects the quality of the timber produced in comparison with what is yielded by crops formed by means of sowing, and that consequently the timber of the former is less able to withstand the attacks of *fungi* later on, is no longer a matter of doubt or question.

'It must, however, be expressly stated that the youthful development of timber crops can afford no reliable indication for the future quality of the mature fall. Expectations, anticipations, and suppositions in this respect have no justification ; as the whole matter depends most essentially on the later treatment of the crops (whether formed by sowing or by planting) during *the operations of thinning out.*'

In estimating the financial advantages likely to accrue from one or other of these methods of formation of crops, the initial costs, of course, form an important factor. And, as planting is, on the whole, more expensive, often very considerably so, than sowing, a choice in favour of the latter method can generally be advised wherever special conditions of soil and climate do not indicate any necessity for planting. Where, however, the inexpensive method of *notching* can be carried out with very young seedlings without any special preparation of the soil, planting operations can frequently be undertaken just as cheaply as, or even cheaper than, sowing ; and in all such exceptional cases planting deserves the preference. But the initial costs of sowing are apt to be mis-calculated; and the subsequent filling in of blanks with transplants often brings up the actual cost to more than would have been the initial expense had planting been chosen as the original method of forming the crop.

Plantations have, in general, notwithstanding the interference and disturbance that takes place before the plants have thoroughly established themselves, a more rapid development in youth, and especially a more energetic growth in height, than young crops resulting from natural regeneration; whilst those formed artificially by sowing occupy a position between these two. These results are explainable by the greater amount of soil-preparation connected with both forms of artificial production and reproduction, and, in the case of plantations, with the larger amount of individual exposure to light, air, and warmth, together with a less prolonged struggle with weeds and rank growth of grass.

Natural reproduction is, on the whole, much cheaper than either of the artificial methods. And, when properly conducted, it affords greater protection against frost, drought, and injurious insects; whilst, at the same time, it maintains the surface-soil subject to the least possible variations in its quality, its covering, and its productive capacity. As greater density of the young crop is usually attained, the thickets grow up remarkably free from branches; but the superior quality of the straight-grained timber is apt to be counterbalanced by the danger to which the young crop is exposed if the poles are allowed to be drawn up too quickly in their mutual struggle for light and air. This struggle for existence is more severe and of longer duration than in the case of crops formed artificially, and especially of plantations; hence careful, oft-repeated clearing and weeding in thickets, and thinnings in pole-forest, are absolutely essential for the avoidance of over-crowding.

For the successful accomplishment of natural regeneration

higher demands are made as regards sylvicultural knowledge
than when total clearance with artificial reproduction is the
method followed; and, when seed years are apt to be irregular,
it often taxes the ingenuity of the manager of large woodlands
to maintain the annual fall of timber without over-stepping
the limits dictated by prudence with regard to the gradual
clearance preliminary to the seed-felling.

It has frequently been objected to natural reproduction that
the growth of the young crop is at first much slower than in
plantations; but such objectors evidently do not attach sufficient weight to the quantitative, the qualitative, and especially
the financial, increment which takes place simultaneously on
the parent standard trees before the final clearance of the
mature crop.

There is one great drawback of plantations, as compared
with crops formed either naturally or by sowing, which seems
inherent from the larger amount of growing-space enjoyed by
each individual plant—that is, an undeniable tendency to
forked or branching growth. This was conclusively proved
by the experimental section of the Forestry Department at
Munich University in 1885 to 1887; and it has been also confirmed for the tropics by my own observations in the extensive
Teak plantations of Burma. The reason is obvious. As the
individual struggle is not at first so great in plantations as
in sowings or in natural regenerations, it more frequently
happens that one or other of the side-buds throws out a shoot
almost equal in size to the axial, true leading-shoot; and as
years advance the fork becomes more developed, thereby
reducing the value of the timber through spoiling the bole
for technical purposes. This tendency may best be observed in
coniferous forests, especially of Spruce, where three, four, and
even five definite and more or less successful efforts at forked
growth are often distinctly recognizable. The same influences
are also at work in other kinds of trees. They tend to dissipate
in ramification the energy in growth which it is the great aim

of the sylviculturist to confine to the main object of the production of a large, clean bole as nearly cylindrical as possible.

The necessity for artificial formation of forests in Scotland during the past century has already been recognized. But as, in very many cases, mature crops are now awaiting utilization and reproduction, the attention of those concerned in the management of wooded estates may well be called to the later advantages offered by natural reproduction, even although at first the rapid increment in plantations seems to point to that method of regeneration as being the most profitable. Yet there is no reason why a compromise should not be effected, natural reproduction being carried out at first to whatever extent it shows itself easy of success, and artificial assistance being then availed of without waiting for the further production of seed by the parent standards. The means thus offered for the formation of mixed crops should in reality add to, rather than detract from, the advantages offered from both a sylvicultural and a financial point of view.

CHAPTER IX

THE TENDING OF WOODS

In one respect there is a strong analogy between human beings and woodland growth; for education or tending is just as necessary in the one case as the other in order to achieve the best results ultimately attainable. With regard to woodland crops this is most particularly requisite in mixed woods. The ultimate shape and value of the boles of the light-demanding species of timber-trees depend to a very great extent on the treatment accorded to the woods during the earlier stages of their development.

Whether the crops be formed naturally by regeneration under parent standards, or artificially by means of sowing or planting, many young woods require a certain amount of protection against dangers arising from inorganic causes like frost, snow, raw winds, and drought, and from organic causes like weeds, insects, and fungoid diseases.

When mature falls of timber are reproduced naturally, or by means of sowing before removing the whole of the mature trees, the requisite degree of shelter can be to a certain extent attained by leaving a sufficient number of parent or other standards per acre. Fortunately, the great majority of trees grown in Britain are of rapid growth during their earliest years, and soon outgrow danger from **frost and drought,** to which the light-demanding genera, Larch, Pine, and Birch especially, and in a less degree also Oak, Ash, and Maples, are naturally less exposed than the shade-bearing species, Beech, Silver Fir, and Spruce. But, when sowings or plantations are formed in the open, it is often necessary to plant out quick-growing and hardy species as **nurses**, in order to minimize any dangers existing from late and early frosts; and for this purpose Birch, Scots Pine, and Larch are best suited, as they naturally suffer little from cold, are rapid in growth, make least demands on the soil, and cast but a light shade around them. In damper localities Willow and Alder are often to be found spontaneously performing similar duties for the benefit of Oak, Ash, and Spruce.

A protection against **raw winds** may be obtained by carrying out the annual falls of mature timber, and consequently the re-formation of young crops, in the direction contrary to that of the prevailing winds. Although this measure is adopted chiefly on account of the older and more valuable crops of timber, to protect them from being thrown, yet the youngest falls at the same time receive the benefit of lying to the lee of the high tree-forest.

That rank growth of **weeds**, whether consisting of grasses or berries, or of woody-fibrous plants like heather, gorse, whortleberries, and the like, must be cleared away so far as they interfere with the development of the young crop, is of course self-evident, unless the future financial value of the mature crop is to be allowed to run the risk of being considerably prejudiced.

It may be remarked here that, as fully explained previously on page 132, coniferous trees are much more liable to such dangers altogether, and' to serious damage from insects and disease especially, than broad-leaved genera, and also that mixed woods suffer less in this respect than pure crops of only one kind of tree practically.

And, in the same way, any outlay for combating the attacks of **injurious insects**, or for annihilating **fungoid diseases** that may have gained a foothold, is just as essential for the present and the future well-being of the crop as if it had actually formed a necessary portion of the initial cost of formation; for no matter whether raised naturally, or artificially by sowing or planting, the development of the young growth is essentially dependent on the time when, and the extent to which, a full and compact leaf-canopy is attained.

Sometimes, in natural regenerations and in sowings, the young seedling growth is so crowded that the crop becomes half choked, and the individual plants, in their resulting struggle for light and air, become drawn up so attenuatedly as not to be able to bear their own weight. In the case of the shade-bearing genera, they then remain to all appearance almost stationary; but the natural process of selection is very much more rapidly effected by the light-demanding kinds of trees. Yet in both classes of trees some artificial aid in the struggle is of untold benefit to the survivors from among which the future mature crop is to be formed. The earlier such operations of tending are begun the better; and when young plants are not required for transplanting, the best and cheapest way is to use the shears as much as possible in cutting off weaklings close to the ground. Should this measure, from one cause or another, have been delayed till the young crops are from ten to twenty years of age, then the general reduction in the number of plants must be undertaken very cautiously, else the remaining saplings, bereft of support, are apt to lean over and even to get quite laid after heavy snowfall. It is therefore best to cut narrow lines through the thicket, thus allowing the plants along the edges a good chance of developing more rapidly and vigorously than those between the lines. These dangers are not so great in plantations, where the greater cost of close planting secure comparative immunity from such a risk.

In practice it much more frequently happens that in young crops, whether formed naturally or artificially, the density is below the normal degree most favourable for their development. Wherever blanks may thus occur, they should be planted up at once with the species of tree best suited for soil and situation, and most likely to be in accordance with the wishes of the proprietor in regard to the ultimate form of the crop.

It also, however, not uncommonly occurs that the general development of the young crop is somewhat backward, in which case it may be necessary to stimulate its energy by the introduction, temporarily, of quick-growing and lightly-foliaged species, not apt to cast any heavy shade around them. In general the special trees chosen for performing this duty are again Scots Pine, Larch, and Birch, on ordinary classes of soil, or Aspen, Willow, and White Alder, on land of a moister description. By planting single rows of these trees at good distances apart, very stimulating effects can be attained in backward growth of young crops of Oak, Beech, Spruce, and Silver Fir. As, however, this measure is temporary only, and not permanent, these rows of supplementary nurses should be gradually removed wherever, and whenever, the growth of the principal genera forming the crop permits of this being done.

When once the normal density of canopy has been attained, the manner in which the different operations of tending are carried out in thickets of sapling growth, in young pole-forests, or in tree-forests, has unmistakable influence on the immediately subsequent development, as well as on the quantity, the quality, and the financial value, of the future mature timber-crop. These operations are of three distinct kinds, viz. :—

1. **Weedings and clearings**, including all operations involving an outlay which is not covered by the amount realizable for the disposal of the material cut out.

2. **Thinnings**, when the costs of the removal of the superfluous or undesirable poles or trees are covered, or more than covered, by their sale.

3. **Partial clearances**, carried out in tree-forest, after the chief growth in height has been completed, with a view to the more rapid development of large-girthed, full-wooded

boles—that is to say, for the production of stems with good top-diameter relatively to the base or butt-end.

1. Weeding and Clearing.

Practically, weeding and clearing, and thinning, are identically similar measures. But in the former case the outlay is properly debitable to the cost of formation of the young crop. The operations are cultural charges which must be met in almost every case, as well as the cost of soil-preparation, sowing, planting, filling up of blanks, &c. As the outlays for weeding and clearing are, as a rule, greater in thickets formed by natural reproduction, or by thick sowing, than in plantations, this tends to equalize the costs of formation, although lower initial outlay is often claimed as an advantage of the former methods over planting. Comparatively few crops are raised from seed without softwoods (Birch, Aspen, Willow) at the same time managing to effect a foothold on the soil, which they often retain with extreme tenacity by throwing up stool-shoots and suckers whenever they are cut back. They thus often interfere with the normal development of the young crop only in just a less degree than if they had been allowed to remain where they took root uninvited. The removal of all such intruders is the special aim of **weeding**; whilst **clearing** is undertaken for the removal of all such individuals, of whatever species, forming part of the original crop as may, if allowed to stand longer, so as to compete with them for light and air, detrimentally affect the further development of other kinds of trees which it is desired to encourage or to stimulate in energy of growth.

So long as the material to be thus removed is gaining in value, the object is not to remove as much as one safely can, but rather to confine operations solely to the removal of what is necessary for the well-being of the rest of the crop. The process of clearing must be repeated as often as necessary; but under ordinary circumstances, in fairly populous tracts, or where small poles are required for pea-sticks, hop-growing, and the like, they should cease to be requisite before the crop enters on the period of its most active growth in height as pole-forest. Under such circumstances **thinnings** should begin early, and may often prove highly remunerative.

From the purely financial point of view the results attainable by **weeding and clearing** are commensurate with the ultimate disadvantages that may probably, and, as continental experience has shown, will almost invariably, accrue, if such cultural measures be neglected. Of these the principal include danger from accumulations of snow and ice, ravages by insects finding a suitable breeding-place in sickly, suppressed, or dominated saplings, and diseases of roots and stems occasioned by fungi like *Trametes, Agaricus*, &c. Neglect of these early measures of tending leads in a very short time to dissipation of the vital energy of the crop, to the overstocking of the area with an excessive number of individual stems of small technical or financial value, to the crowding out and suppression of the more valuable trees by kinds of less value, and ultimately to the necessity of clearance and utilization of the crop at a date much earlier than its technical and commercial maturity would otherwise have been indicated financially.

Wherever reliable data may exist for a comparison of the ultimate returns obtained from two different crops grown on similar soils and situations, and both formed by similar methods, though subjected to essentially different treatment during the early stages of growth, it will invariably be found that, when a less valuable species or individual has been allowed to interfere with the development of a more valuable, on comparison of the net ultimate returns from the mature crop plus all the intermediate returns from thinnings, capital-ized up to the time of utilization of the two mature falls of timber, then the capitalized outlay on weedings and clearings can be proved to have been money very well laid out simply as a further investment of capital.

2. Thinnings in Pole-forest and Young Tree-forest.

Under **thinnings** are understood all operations for the removal of unnecessary species or individuals in timber crops, dating from the time that the outlay thus occasioned is covered by the returns obtained for the material removed, and thence onwards throughout the pole-forest and the tree-forest stages of growth, till a commencement is made of the clearances for reproduction or utilization, or until **partial clearances** are made, after the chief growth in height has been completed, for the more rapid development of large-girthed boles with good top-diameter.

The ultimate quantitative and qualitative outturn from woodland crops depends much more on the proper conduct of the thinnings than on the preliminary operations of weeding and clearing. Practically, there is generally a period, of shorter or longer duration, in which the thickets formed from natural reproduction or from sowing are almost impenetrable ; and in mixed forests, in which the more valuable subordinate species occur individually, and not in patches, this is often the most critical period for these latter kinds of trees.

The object of thinning any crop is to stimulate and to assist its development so that the aim of the proprietor may be attained as completely and as soon as possible, without necessitating any such interruption of the canopy as may prejudice either the growth in height of the crop or the productive capacity of the soil. The crops which stand most in need of thinning, and of a frequent repetition of the operation, are those in which the individual stems are all of about the same age ; as the natural struggle for individual existence is then more keen and prolonged than where some plants of advance-growth have from the very outset won an advantage. In this process of natural selection four classes of stems become distinguishable, viz.

(1) *predominating*, (2) *dominant*, (3) *dominated*, and (4) *suppressed*.

As, during the gradual process of death and decay that takes place when stems become suppressed, certain physiological conditions obtain, which are extremely favourable to the breeding of insect enemies and the propagation of fungoid diseases, this class of individual stems should invariably be removed during thinning operations. By this means a freer circulation of air and a better utilization of the soil-nutrients are also at the same time effected.

To what extent a simultaneous removal of the dominated class should proceed, depends in each case on the concrete factors of soil, situation, species of tree, age of crop, mode of treatment, &c. If the struggle were left to nature unassisted it would be carried out most rapidly on fertile soils; but, on soils of merely average or inferior quality, it would be so much prolonged in the case of shade-bearing species of trees as to result in overcrowding of the crop throughout the pole-forest and the young tree-forest stages of growth.

From the sylvicultural point of view, however, the main objects of care are the timber-trees which it is desired to utilize on their attaining full financial maturity; and any measures that will aid in stimulating them to the speedy attainment of large remunerative dimensions must deserve favourable consideration. Year by year the number of individual stems, from which it is possible that the future mature crop may be formed, becomes diminished; and the removal and the utilization of all unnecessary stems are not only a means of improving the growth of the predominating and dominant portions of the crop, but are likewise a source of revenue, which should be realized as soon as available, in order to reduce the capital cost of the crop both at present and up to its final clearance. And these preliminary or intermediate returns are often of no mean value where any good market exists for small timber like poles and pit-props. Thus Grebe, a dis-

tinguished sylviculturist of Thuringia, who was for many years
the Principal of the Forest School at Eisenach, wrote as
follows [1] :—

'Granted that the thinnings begin at the proper time—in the case of
the Beech about the twenty-fifth to thirty-fifth year, the Spruce about the
twentieth to thirtieth year, and the Scots Pine about the fifteenth to
twenty-fifth year—that they are regularly conducted, that the yield
therefrom is not reduced by any peculiar local circumstances (such as
right of collection of windfall, or interruption of density in canopy due to
snow accumulations, &c.), it may be expected that on the average the
proportion which the intermediate yield from thinnings bears to the final
yield of the mature crop will be :—

In *Beech Woods*, with a rotation of—

> 80 years, from 12–20 per cent.
> 100   ,,   ,,   14–25   ,,
> 120   ,,   ,,   16–30   ,,

The lower percentage is obtainable from poor soils, the higher from
the better classes of soil.

In *Spruce Woods*, with a rotation of—

> 60 years, from 15–17 per cent.
> 80   ,,   ,,   20–22   ,,
> 100   ,,   ,,   23–26   ,,

The higher percentages are obtainable from the inferior classes of soil.

In *Scots Pine Woods*, with a rotation of—

> 60 years, from 18–24 per cent.
> 80   ,,   ,,   22–28   ,,
> 100   ,,   ,, about 25   ,,

The higher percentages are obtainable from the inferior classes of soil.'

For pure Spruce forests on the Harz Mountains, Theodor
Hartig found the following total number of stems per acre [2] :—

At	20 years of age,	9,265 [3]	of which	49 %	
,,	40 ,,	1,249	,,	42	might be advantage-
,,	60 ,,	603	,,	32	ously removed by
,,	80 ,,	388	,,	21	thinning.
,,	100 ,,	282	,,	11	
,,	120 ,,	238	,,	4	

[1] *Die Betriebs- und Ertragsregelung der Forste*, 1879, p. 300.
[2] Gayer, *Waldbau*, 3rd edit. 1889, p. 15.
[3] This only refers to natural reproduction and sowing; it would of course
be much less in the case of plantations. But compare with the above the
remarks made on page 57.

In mixed forests the practice of thinning must, of course, be conducted on quite a different principle from that obtaining with regard to pure woods. Thus in order to save a valuable species of tree, it may very often be found necessary to cut out other kinds of more energetic growth, which may threaten to prejudice its development, or even to endanger its existence.

As regards the influence which the species of tree has on the extent to which thinnings are necessary, Schuberg[1] found in the Black Forest that in forty to eighty-year-old crops, which had been regularly thinned, the following were the results on soils of average quality :—

Kind of Crop.	Scots Pine.	Spruce.	Beech.	Silver Fir.
Average number of stems per acre	545	619	686	864
Absolute individual growing - space in square feet	80	70	63	50
Relative individual growing-space	100	87	79	63

If the chief light-demanding genera (Pine, Larch, Oak, Ash, Maple, Birch, &c.) be grouped together, and be compared with the various shade-bearing kinds of trees, it will be found that on the whole these latter only require on the average from 50 to 75 %, or one-half to three-quarters, of the growing-space demanded by the former for their proper development as a woodland crop.

The quality of the soil is also a very important factor in determining the number of stems per acre. Owing to the more

[1] Gayer, *op. cit.*, p. 550.

equal strength with which the individual struggle for supremacy is waged on soils of indifferent quality, the growing-space per stem is, up to the sixtieth to eightieth year, often considerably less than on better soils, although in the case of the Scots Pine this difference is less marked than in any other species. The practical lesson thereby conveyed is that, on soils just of or below the average in quality, both financial and purely sylvicultural considerations point to the advisability of stimulating, by means of early, oft-repeated, and moderate thinnings, the development of all individuals which it is desired to retain. **Thin early, often, and moderately** is the golden rule of tending woodlands. Anything approaching interruption of the canopy during the period of most active growth in height must of necessity lead to a further development of the crown at the expense of the length and cleanness of the bole ; and this will consequently tend towards a diminished technical and monetary value of the timber produced.

Between the two extremes of overcrowding and of free individual growing-space there must be in every kind of crop a happy medial point or a **normal density of canopy**, at which the quantity and the quality of the crop attain their combined maximum. To show the practical effect which thinning has on the increment of the basal area of the individual stems left forming the crop, the following example may be given. In two different parts of an extensive forest three areas were selected on soil of similar nature and quality, and thinnings were carried out on equal areas in the following degrees :—

I. *Slight*, only dead and dying poles being removed.

II. *Moderate*, all suppressed poles being removed.

III. *Heavy*, all dominated poles being removed.

The conditions as to soil and situation were in each locality as similar as possible, the soil being a fresh, loamy sand throughout. The results of the experiment were tested five years after the thinnings took place.

I.—In the first instance, that of a thirty-seven-year-old crop of Spruce, raised by planting, the data[1] were :—

Previous to thinning :

Area No.	Nature of thinning.	No. of stems per acre.	Total Basal area, sq. ft.
I.	Slight	2,126	242
II.	Moderate . . .	2,100	251
III.	Heavy	2,044	234

During the thinning there were removed—

Area No.	Stems per acre.	Basal area removed, sq. ft.	Percentage of Basal area removed.
I.	502	9	About $3\frac{3}{4}$
II.	822	28	,, 11
III.	971	37	,, 16

In consequence of the thinning, *the arithmetical mean diameter at breast-height* ($4\frac{1}{3}$ ft.) of the stems that were left was found to be—

On area I, 5·12 inches ;

On area II, 5·68 inches ;

On area III, 5·84 inches ;

whilst five years later, just before another thinning was to take place, it was found to have increased—

On area I to 5·48 inches, or an average diametral increment of 0·36 inches per stem.

---

[1] These and the following data are, by permission, excerpted from Dr. M. Behringer's prize thesis *Ueber den Einfluss wirthschaftlicher Massregeln auf Zuwachsverhältnisse und Rentabilität der Waldwirthschaft*, 1891, pp. 19 et seq.

On area II to 6·16 inches, or an average diametral increment of 0·48 inches per stem.

On area III to 6·40 inches, or an average diametral increment of 0·56 inches per stem.

II.—In the other locality, a thirty-six-year-old Spruce wood, raised by sowing, the data were :—

Previous to thinning :

Area No.	Nature of thinning.	No of stems per acre.	Total Basal area, sq. ft.
I.	Slight	7,707	219
II.	Moderate . . .	7,725	235
III.	Heavy	7,497	220

The operation of thinning removed—

Area No.	Stems per acre.	Basal area removed, sq. ft.	Percentage of Basal area removed.
I.	4,736	29	About 13
II.	5,619	53	,,　22
III.	5,915	71	,,　32

when *the arithmetical mean diameter at breast-height* of the remaining stems was found to be—

On area I, 3·44 inches ;

On area II, 4·00 inches ;

On area III, 4·20 inches ;

and, *just previous to another thinning five years later*, it had risen—

On area I to 3·88 inches, or an average diametral increment of 0·44 inches.

On area II to 4·48 inches, or an average diametral increment of 0·48 inches.

On area III to 4·72 inches, or an average diametral increment of 0·52 inches.

The actual average increment in basal area was found to be per annum per acre—

I. In the Plantation.			II. In the Sowing.		
		sq. ft. 6·6			sq. ft. 8·9
Area	I.		Area.	I.	
„	II.	7·7	„	II.	8·7
„	III.	8·2	„	III.	8·2

Comparing the average annual increment with the total basal area at the middle of the period of five years, the percentage of annual basal increment was found to be—

In the Plantation.			In the Sowing.		
		per cent. 2·68			per cent. 4·27
Area.	I.		Area.	I.	
„	II.	3·22	„	II.	4·39
„	III.	3·85	„	III.	5·02

The obvious conclusions from these data seem to be that in plantations, in which from the very outset each individual plant has a fixed amount of growing-space, the natural process of selection proceeds so much more rapidly than otherwise that—

I. There is less risk of any overcrowding of the leaf-canopy.

II. Despite the smaller number of stems per acre, the total

basal area (i.e. the sum-total of all the stem-superficies at breast-height) is certainly not less than in dense crops raised from seed or by natural regeneration.

III. A far less percentage of the total number of stems belongs to the dominated and suppressed classes.

IV. The stimulation of the increment in basal area takes place in accordance with (though it is not necessarily proportional to) the degree to which the thinning takes place.

So long, therefore, as thinning is not carried to such an extent as to prejudice the energy of growth in young crops that have not yet nearly completed their activity, i.e. which have not yet culminated in average annual increment, a free thinning stimulates to the earlier maturity of the crop, and is therefore decidedly advantageous from a financial point of view. Thinnings are, generally speaking, said to be **slight** when only dead or dying poles are removed, **moderate** when all suppressed poles, and also a portion of the dominated poles, are cut out, and **heavy** when at the same time all the dominated poles are eliminated[1].

It may be noted here that, *in tree-forest*, a slight thinning removes about 5 % of the basal area of the stems at breast-height in crops of normal density of leaf-canopy, a moderate thinning about 10 %, and a heavy thinning about 15 %. Where it goes beyond this last degree, it becomes in reality a partial clearance, and must be regarded as such.

The average percentage of basal increment has been given above; but it must be observed that these figures are subject to modification. For the actual percentage of basal increment is

---

[1] In the special case referred to on p. 64 the *moderate thinning* was particularly stated not to include any of the dominated stems, as these were all included in the *heavy thinning*. But it must be borne carefully in mind that the thinnings were carried out in dense crops of Spruce, which is one of the most *shade-bearing* kinds of trees. For crops of *light-demanding* trees, like Oak, Larch, Pines, Ash, Maple, Sycamore, Elm, Birch, Willow, and Poplar, even a *moderate thinning* would certainly remove a portion of the dominated individuals; but the extent to which these should be cut out would, in each concrete case, depend on the kind of trees forming the crop, the general vigour of the latter, and the nature of the soil and situation.

greatest just immediately after the thinning takes place ; and it gradually decreases as the crowns close up again towards the termination of the period at which another thinning becomes necessary. The physiological explanation of this decrease is to be found in the fact that the assimilative functions of the lower foliage again become weakened through the gradual exclusion of the light, air, and warmth requisite for assimilation.

Experience has shown that the height of timber crops, as well as being practically proportional to the quality of the soil for the species of tree in question, is, in all regular crops growing on soils and situations of similar quality, and under similar conditions as to their development, as a rule proportional to their basal area until they have completed their chief growth in height.   But when the conditions of development vary, as must be the case when, *ceteris paribus*, certain woods are only slightly thinned, and others moderately, or perhaps even heavily thinned, then the natural proportions between height and girth become to a greater or less extent interfered with.   The fuller the leaf-canopy remains, the more does the form of the bole approximate to the cylindrical (i.e. the higher does the *form-factor* become) ; and the freer the growing-space, the greater is the tendency to conical growth of bole (i.e. the lower does the *form-factor* become) with simultaneous increase of branch development.   Experiments carried out in Beech and Scots Pine woods by Dr. Behringer, under the direction of the experimental section of the Forest Branch of Munich University, went to prove[1] that—

' When thinnings were carried out freely, the development in height was relatively much greater than the diametral increment : and that at any rate during a certain period in the growth of crops, the heaviest degree of thinning produced the loftiest and the cleanest boles.'

Of course, the correctness of this statement, or of any other dictum with regard to most of the operations of practical sylviculture, depends in each case on the nature and condition

[1] *Op. cit.*, p. 28.

of the crop in question as to species of tree, age and mode of formation of crop, condition previous to, and at time of, thinning, nature of soil and situation, &c.　But Weise[1] seems perfectly correct in making the broad generalization that —

' When a crop is too thick, increment in height and in girth both suffer to an extent not compensated by the outturn yielded from the greater number of stems ; when it is too thin, however, the crop remains backward in growth in height.'

Experiments made in order to endeavour to formulate a natural law regarding the effect of slight, of moderate, and of heavy thinnings on increment in height, girth, form-factor, and total yield of timber per unit of area, have hitherto failed to yield any practical result.　On the whole they simply lead one in a general way back to appreciate the wisdom and the practical value of the rule, founded on experience, that '*thinnings should be begun early, carried out moderately, and repeated frequently.*'　This does not militate against the correctness of the theory and the practice of partial clearance, after once the chief growth in height is completed, for the purpose of stimulating the trees to rapid increment in girth and to improvement in the shape of the bole.

And, practically, the same conclusion must be come to when the matter is viewed from the financial standpoint.　Although increased returns from material thinned out and profitably disposed of—regarding them not only as items reducing the capital represented by the growing stock, but also as sums which, for correct reckoning, must be taken as growing in value at compound interest till the whole mature crop is harvested— would point to the desirability of heavy thinnings, and would perhaps yield fair results where the market for small timber is favourable, yet the future well-being of the ultimate crop is undoubtedly the main object to be kept in view.　And here, again, experience has shown that considerations affecting both the productive capacity of the soil, and the normal development

---

[1] See *Chronik*, 1881, p. 25.

of the crop, are most favourably kept in view when the degree to which thinning is carried out, at each time of repeating the operation, is only moderate—although on soils of better than a merely average quality it may well be somewhat freer than on those of a poorer nature.

3. Partial Clearances in Tree-forest.

Partial clearances are made in tree-forest when over 15 % of the basal area is removed for the express purpose of affording freer supplies of light and air, so as to stimulate the rate of increment, and thus enhance the technical and the financial value of the timber.

This sylvicultural operation, which is merely a more vigorous and emphasized expression of the theory of thinning, may be carried out after once the timber crop has passed through the most active period of growth in height; for then the advantages, offered by increment in girth and in full-woodedness towards the top-end of the bole, are not, to anything like the same degree as during any earlier stage of development, prejudiced by such increment being perhaps stimulated at the expense of the rate of growth in height. The removal of up to 15 % of the total basal area, when the crop stands in full canopy, is still a thinning; whereas the removal of 18 or 20 % is already a partial clearance. But the latter operation is practically merely a continuation of the former; only it is carried out with a freer hand.

Whilst, in thinnings, the conservation of the productive capacity of the soil is one of the most important objects to receive consideration, in a partial clearance its maintenance, by means of the tree-crop alone, is seldom attainable except on soils of the very best quality; hence underplanting is usually necessary, simultaneously with the partial clearance being made [1]. Throughout the whole of the operations of tending—

---

[1] A fuller consideration will be given to this subject in Chapter XI, *On Underplanting.*

the weedings and clearings, the thinnings, and the partial clearances—the same object is definitely kept in view, viz. the best possible ultimate development of the individual stems forming the chief financial factors in the mature crop. During all these tending operations the more valuable kinds of trees are not only protected against other species threatening their existence, but efforts are also consistently made to procure for them, at all the various stages of their development, such conditions relative to growing-space, &c., as may produce the greatest outturn of long-stemmed, large-girthed, and full-wooded timber in the shortest space of time—or, in other words, as may produce the most valuable timber, technically and financially, at the least cost of production.

In practical forestry this method of partial clearance is to a great extent naturally confined to Oak, Scots Pine, and Larch ; for the other light-demanding species of trees, Maples, Elm, Ash, Birch, &c., which are in request for ornamental work chiefly, are generally utilized during the later thinnings. When forests, in which the ruling species is Silver Fir or Beech, are to be naturally reproduced, the same stimulation to increment in girth, and to improvement in the shape of the bole, is practically obtained during the clearances made preparatory to, and following after, the seed-felling. In sheltered localities the effects of partial clearance are also very profitable in the case of Spruce crops, which, owing to the great danger to which this shallow-rooting species is exposed from windfall, are not generally regenerated naturally, except under very favourable circumstances as regards protection from winds.

When the first partial clearance is made—in Oak woods about the fiftieth to sixtieth year, in Scots Pine between the thirtieth to fiftieth year, and in Larch about the thirtieth to thirty-fifth year—the fall is confined to all stems of inferior development, or to such as do not give promise of ultimately producing timber of the better class. Later on, in about five, or ten, or fifteen years, when the standard trees gradually begin to approach each other in an endeavour to form canopy, the clearance has of course ·to be repeated, in order that the foliage on the lower branches, and within the interior of the crowns, may not be hindered in its assimilative functions, owing to the decreased measure of exposure to light and air to which they must be reduced in anything like close canopy. As in regard to thinnings, there can be no hard and fast rules for the conduct of partial clearances; for so much depends on the quality of the soil and situation. But, practically, they should be repeated at intervals of about five to ten years at first, and about ten to fifteen years later on. **Oak woods** treated in this manner about the fiftieth to sixtieth year should, according to Gayer[1], yield a quantitative annual increment of from 3 to $3\frac{1}{2}$ % up to their 100th year, and 2 to $2\frac{1}{2}$ % after that, without taking into consideration the qualitative, technical, and financial increment simultaneously rising more rapidly. At 120 years of age the Oak woods treated thus should show about thirty-six to forty-eight stems per acre. **Larch crops** partially cleared from the thirtieth to thirty-fifth year onwards should, on situations naturally adapted for their growth, at sixty to seventy years of age show about sixty to seventy trees per acre with an annual increment of from 3 to 4 %. In **Scots Pine woods** a repetition of the process is only advisable on the better classes of soil, as, otherwise, experience has shown that better results can generally be achieved by letting the whole of the standard trees come together to form the light canopy which is their characteristic on inferior soils and situations, and beneath which a crop of underwood can usually thrive fairly well. In **mixed woods of Spruce, Silver Fir, and Beech** excellent results have been attained by beginning to thin rather heavily about the thirtieth year, repeating the thinnings every tenth year, and then making a partial clearance between the sixtieth to seventieth year, so as to leave per acre about 120 to 160 trees of sound, promising growth. In crops thus

---

[1] *Op. cit.*, pp. 574, 575.

treated, after the lapse of twenty years natural regeneration will generally be found to have been effected spontaneously, the standard trees will be of good marketable dimensions and excellent quality, and the seedling growth will vary from about 3 to 15 feet in height.   Blanks left by the removal of the standards can easily be most advantageously filled up by planting with Pine, Oak, Larch, Ash, Maple, &c.

Of course, an essential condition for the success of this measure is that the crops are still in active vitality and capable of being stimulated to quantitative and qualitative increments. The operation would be useless in the case of trees already entering, even prematurely from injudicious treatment or any other cause, into the stage of senile decay.   With reference to the relative and the absolute increment on individual stems after partial clearance, Behringer[1] gives some interesting and instructive data, which are exhibited in tabular form on the two following pages.

Measurements made at the same time showed that the rate of growth in height had been sensibly diminished in consequence of the greater amount of growing-space afforded after the partial clearance.   But as the operation was not performed until after the most active period of growth in height had been completed, this diminution was of comparatively little technical or financial importance.

So far as these experiments go they prove not only that, owing to the larger growing-space, each of the standards was stimulated to produce more than double the increment in cubic contents which it had before the partial clearance took place, but also that, whilst in regard to the Spruce the enhanced increment culminated within the first decade after this operation, the culmination was delayed in the cases of the Scots Pine and the Silver Fir until the second decade had been entered on.   This had also previously been found by König to be the case in Beech woods.

<hr>

[1] *Op. cit.*, pp. 59, 60.

Percentage of Superficial Increment at various sections of the Stem.

Stem Sections 13·3 feet apart.	Spruce.				Scots Pine.				Silver Fir.			
	Before		After		Before		After		Before		After	
	Partial clearance at Age (in Years).				Partial clearance at Age (in Years).				Partial clearance at Age (in Years).			
	58–68.	68–78.	78–88.	88–90.	56–66.	66–76.	76–86.	86–90.	54–64.	64–74.	74–84.	84–90.
At felling height	2·23	1·91	4·44	2·22	2·74	2·19	3·76	3·08	3·85	2·06	4·38	6·26
13·3 feet above felling height	3·07	1·88	4·66	2·08	1·96	2·64	3·26	3·56	4·17	2·34	2·81	4·10
26·6 ,, ,, .	3·61	2·16	4·21	2·68	2·90	2·47	3·13	4·19	7·35	2·71	3·30	4·27
40·0 ,, ,, .	6·35	2·95	5·06	2·63	6·96	2·90	3·83	4·56	15·50	4·36	3·45	4·44
53·3 ,, ,, .	15·7	5·40	4·88	3·11	12·30	3·54	4·30	3·00	—	10·50	8·50	2·89
66·6 ,, ,, .	—	18·60	9·30	3·01	—	6·33	7·23	3·60	—	—	—	7·60

Percentage of Increment in Cubic Contents during each period.

—	4·27	2·61	4·74	2·50	4·47	2·84	3·78	3·72	7·17	3·35	4·10	4·60

Percentage of Superficial Increment at breast-height during the first 10 years after the partial clearance.

—	—	—	4·43	—	—	—	3·68	—	—	—	4·02	—

Stem Sections 13·3 feet apart.	The Relative Superficial Increment of the various sections—that at felling height for the 10 years previous to the partial clearance being = 1·00.										
	Spruce.				Scots Pine.				Silver Fir.		
	Before		After		Before		After		Before		After
	Partial clearance at Age (in Years).				Partial clearance at Age (in Years).				Partial clearance at Age (in Years)		
	58–68.	68–78.	78–88.	88–90.	56–66.	66–76.	76–86.	86–90.	54–64.	64–74.	74–84.	84–90.
At felling height	0·95	1·00	2·77	2·01	0·98	1·00	2·10	2·35	1·35	1·00	2·56	5·27
13·3 feet above felling height	1·05	0·84	2·46	1·61	0·56	0·90	1·41	2·04	1·19	0·95	1·41	2·63
26·6 ,, ,, .	0·99	0·80	1·91	1·73	0·63	0·69	1·09	1·91	1·31	0·84	1·30	2·22
40·0 ,, ,, .	1·00	0·76	1·60	1·32	0·97	0·69	1·17	1·93	1·27	0·91	1·03	1·79
53·3 ,, ,, .	0·79	0·70	0·95	0·80	0·84	0·54	0·88	0·88	0·72	0·80	1·32	0·84
66·6 ,, ,, .	c·17	0·32	0·46	0·29	0·46	0·52	0·61	0·52	—	—	0·56	0·57

Annual average Increment in Cubic Contents during each of the different periods.

| In Cubic Feet . . . . | 0·3744 | 0·4464 | 1·3752 | 0·8568 | 0·8784 | 0·8064 | 1·3788 | 1·8720 | 0·7308 | 0·5868 | 0·9576 | 1·5156 |

Relative Increment in Cubic Contents, the volume of the tree during the period of 10 years previous to the partial clearance being = 1·00.

| — | 1·15 | 1·00 | 2·84 | 1·78 | 1·09 | 1·00 | 1·71 | 2·32 | 1·25 | 1·00 | 1·63 | 2·58 |

That, at the same time, a qualitative enhancement takes place along with the quantitative increment of the timber is clear from the researches of R. Hartig, who asserts[1] that in general, without variations according to the species of tree, the quality of the timber grown in the comparatively full exposure to light and air is better than that formed under other conditions. This is due to the fact that the extent to which transpiration is carried on through the foliage does not increase *pari passu* with the enhanced rate of increment, as the conductive capacity of the vessels in the new zones of wood is relatively smaller, in consequence of the ligneous deposits within the cellular tissue being thicker and denser. And, further, when underplanting is simultaneously carried out, the undergrowth keeps the soil and the root-system cool, and thus retards, for at least a fortnight, the awakening of active vegetation in spring; hence the latter is only entered upon when the weather has become somewhat warmer, and when the assimilative and constructive processes can consequently be performed more thoroughly and energetically, owing to their taking place under more favourable conditions as to warmth and sunshine.

To determine whether or not a partial clearance is likely to be financially remunerative, practical figures, based on past average results for crops of similar kinds on soils of similar quality, are requisite. Where these are available, the reckoning is very simple by means of the following equation :—

$$\mathrm{CV} + t_a\,v_a + \ldots\ldots t_g\,v_g = (\mathrm{C_1V_1} + t_a\,v_a + \ldots\ldots t_\zeta\,v_\zeta) + x$$

When C = the cubic contents that may be expected from the mature crops with ordinary treatment only,

V = the value of same per load, or per cubic foot, at time of harvesting the mature crop,

$t_a \ldots t_g$ = outturn in cubic contents from ordinary thinnings from now till harvesting of the mature crop,

$v_a \ldots v_g$ = the sale-price of same per load, or per cubic foot, capitalized until the clearance of the mature crop.

[1] *Centralblatt für das gesammte Forstwesen*, 1888, pp. 8, 363.

And C_1 = the cubic contents expected from mature crop after treatment by partial clearances,

V_1 = the value of same per load, or per cubic foot, at time of final clearance,

$t_a \ldots t_\zeta$ = outturn in cubic contents from partial clearances up till the harvesting of the crop,

$v_a \ldots v_\zeta$ = the sale-price of same per load, or per cubic foot, capitalized until clearance of the mature crop takes place.

When x is a minus quantity, then the method of partial clearance is remunerative, and therefore advisable.

,, $x = 0$, then there is no financial advantage to be gained by partial clearances.

,, $x = a$ plus quantity, then the ordinary method of treatment by thinnings merely is the more advantageous.

But when underwood, which should almost always be formed—even on the better classes of soil—can only be produced after a certain amount of outlay, then the complete and correct formula for estimating the financial prospects of the operations must be :—

$$CV + t_a\, v_a + \ldots t_q\, v_q \underset{>}{\overset{<}{=}} C_1 V_1 + t_a\, v_a + \ldots t_\zeta\, v_\zeta - \mu + r$$

or, in another form similar to that originally given,—

$$CV + t_a\, v_a + \ldots t_q\, v_q = ((C_1 V_1 + t_a\, v_a + \ldots t_\zeta\, v_\zeta) - \mu + r) + x$$

Where μ = the outlay for producing underwood, capitalized until final clearance of the mature crop,

and r = the market value of the underwood at the time of clearance of the mature crop.

Practical experience in Germany has shown that this method of partial clearance should, under ordinary circumstances, only be adopted on the better classes of soil, and when undergrowth can be formed either naturally or without any considerable outlay.

Pruning or Removal of Branches.

High-forest crops, which have been properly tended and thinned throughout the various stages of their growth, require, as a rule, no clearance of branches in order to improve the shape of the bole. The cutting and the lopping of branches are both merely measures adopted, often at considerable outlay, to cover the effects of faulty treatment of the crop during the past, and are, in general, not requisite where trees have gradually been accustomed to a fuller measure of individual growing-space. In the case of copse or standards over coppice some diminution in the amount of shade cast may be necessary for the well-being of the underwood; and, within certain limits, this can be effected by thinning the crowns of the standard trees forming the overwood.

Under ordinary circumstances, the removal of living branches can only take place to a limited degree, and must in any case (for financial reasons) be confined to the more valuable species of trees. The manner in which the crown is set on the tree is of well-known influence in determining the shape of the bole, as stems with lofty crowns are more full-wooded at the top-end, and approximate more to the cylinder in shape, than such as have deep-seated crowns of foliage. In order that any benefit may be derived from the operation, it is necessary that pruning should be carried out when the trees are in their full energy of growth, and as early as possible before the period fixed for their mature fall.

In many cases, too, where the crowns of the trees are very dense, a fillip may be given to their general energy in growth by thinning out some of the branches, so that the foliage remaining may have a larger share of the nutrients extracted by the root-system from the soil. The stimulation thus given to the energy of assimilation is perhaps nowhere more noticeable than with regard to young Larch trees, the lower portions of whose foliage may have already been attacked by the Larch mining-moth (*Coleophora laricella*).

The removal of dead branches, when carefully conducted, so as to leave no ragged surface likely to offer a favourable germinating-bed to fungoid spores, interferes in no way with the performance of the vital functions of the rest of the tree. On the contrary, when the operation is carried out smoothly and close to the stem, it prevents the formation of hard, horny knots in the wood; and this consequently enhances the value of the timber, considerably diminishes the danger from fungoid diseases, and favours an early and complete cicatrization of the wound-surface. It is therefore particularly to be recommended in the case of Oak, Pine, Larch, Spruce, and Silver Fir, whose dead branches are much more apt than in other genera to form hard, tough snags, which, becoming gradually embedded in the stem by the new growth of annual zones of wood, diminish the value of the timber for technical purposes, and thereby affect its market price.

The removal of living branches is, however, a direct interference with the vital condition of the growing tree, and can therefore only be ventured on to a limited extent. Under no circumstances should more than one-third of the total quantity of foliage be removed at one time; practically, only about one-fifth of the foliage is generally the extent removed during such operations. Experience has shown[1] that, even in the case of quick-growing kinds of trees, the removal of living branches should ordinarily be confined to those that are not over $2\frac{1}{2}$ inches in diameter for conifers, or not over 4 inches for broad-leaved species of trees, as, otherwise, the process of cicatrization cannot take place quickly enough to ensure hindrance of the germination of fungoid spores producing disease.

As the broad-leaved species of trees are not so liable as conifers to fungous infection, they usually stand trimming better. With reference to this matter Professor R. Hartig remarks that[2] :—

'When branches are removed from a tree, it thereby acquires a predisposition for a number of wound-diseases, of an infectious or non-

---

[1] See Ney, *Die Lehre vom Waldbau*, 1885, p. 300; also Gayer, *op. cit.*, p. 584.
[2] *Lehrbuch der Baumkrankheiten*, 1889, p. 15.

infectious nature, which may be obviated by timely and effective closure, i. e. by antiseptic treatment.'

And at another place he adds[1] :—

' In broad-leaved trees, especially the Oak, to which my investigations have hitherto been confined, wounds over 4 to 4.8 inches in diameter, should not be made.'

The species of trees which best stand the removal of green branches are the Oak, Silver Fir, Larch, and Pines, when they are in energetic growth on favourable soil and situation and have normally developed crowns. Ash, too, can well bear the operation ; but, as Ash and Alder both grow with naturally clean stems, pruning is hardly requisite in their case. Whenever the removal of branches takes place to any great extent from Oak, Maple, Sycamore, or Elm, the result is that the bole has a tendency to become covered with shoots from the dormant buds. This, of course, interferes very considerably with the technical and financial value of the stem, besides, particularly in the case of the Oak, exposing the tree to the danger of becoming '*stag-headed*' or dead in the upper portion of the crown. The softwoods—Birch, Willows, and Poplars—are not naturally adapted for this kind of treatment, owing to the comparatively slight resistance which their soft, porous wood is able to offer to the development of fungoid disease ; hence any wound-surfaces formed afford only too favourable a germinating-bed for the disease-producing spores.

In order to diminish the danger from fungoid diseases, it is highly recommendable that the wounds should be coated over with some antiseptic substance impervious to moisture. In conifers this takes place naturally by the oozing out of resin ; but some coating must be provided artificially in the case of the broad-leaved genera. The tree-wax, formerly in general demand for the purpose, consisted of a mixture of 1.20 parts (by weight) of bees-wax, 2.70 parts of pure resin, 0.60 of turpentine, 0.15 of wood-oil, and 0.15 of suet, all dissolved in

[1] *Ibid.* p. 228.

warm methylated spirits. Nowadays, however, the simpler method of coating over the surfaces with coal-tar, slightly thinned with oil of turpentine, finds general adoption owing to its cheapness. The coating of tar will only bite into the surface satisfactorily whilst the wood is relatively free from sap; hence it should be administered either towards the end of October, immediately after the vital functions of the foliage have been completed, or at any rate some time during the first half of winter. Unless the removal of the branches takes place in autumn or early in winter, the experiments made by Professor R. Hartig of Munich tend to show that, in addition to the tar not obtaining a good hold on the wound-surface, the condition of the woody tissue is such as to make it less able to resist the penetration of fungoid, disease-producing spores. Practically, the same principle holds good here as in regard to the thinning of young tree-forest; for it is much better to thin out branches to a moderate extent only, and then repeat the operation subsequently if necessary, than to endeavour to effect the object in view by one *coup de main.*

The cost involved by the operation varies, in each particular case, according to the local rate of wages and the handiness of the workmen, the species of trees, the height at, and the extent to, which the branches are to be removed, &c.; but in Germany it has been found to vary generally in amount from about $\frac{1}{4}d.$ to $2d.$ per stem for sawing off the branches and tarring the wound-surfaces. In Britain, it will, for various reasons, probably cost quite the double of this amount; but, even then, it may, under certain circumstances, prove a highly remunerative operation if judiciously conducted.

With regard to the remunerativeness of the removal of branches, Alers, who invented the patent long-handled saws now usually employed in the operation, estimated that if begun in the thirtieth year of age of the crop (coniferous?), and repeated every five years till the fiftieth year, at a cost of about $1\frac{1}{4}d.$ per stem, the net results at eighty years of age, after deducting the various outlays capitalized at compound interest up to that date, would show a profit of 12 %. It is impossible to criticize such a vague estimate without examining all the data closely, and ascertaining what the results would be for a fall of timber fixed at 100, 120, or 150 years; but, at any rate, it is worth remarking that the practical effect of Alers' agitation has been to induce the Prussian Government to undertake regularly the removal of all dead branches in forests grown for the production of valuable timber stems, and not intended merely to be utilized as fuel. From the national-economic point of view this may be advisable in State forests, though for private landowners it is a purely financial question, concerning which local experience will best guide the sylvicul-turist to a sound judgement about the matter. In accordance with the natural laws of tree-growth, as increment begins at the top and is gradually continued towards the base, the removal of a portion of the foliage diminishes the amount of elaborated nourishment available for structural purposes towards the butt-end of the trunk, and thus leads to improvement in the form-factor of the bole; whilst at the same time, as R. Hartig has also shown, the quality of the timber produced improves the more, the less difference there is in the quantitative incre-ment proportionately to the surface of the foliage through which the process of transpiration is carried on. That, in consequence of this artificial thinning of the crown, the growth in height is stimulated, is a point about which experts have hitherto differed, and which has not yet been determined by any series of authoritative experiments.

Thus, while special circumstances have most influence in each particular case relative to the remunerativeness of such measures in private woodlands, the general statement can at any rate be made, that the removal of sickly, rotting branches— or of all those infected with parasitic growth of Loranthus or Mistletoe, or of such as exhibit deformities like twig-clusters (e. g. those caused by *Aecidium elatinum* on Silver Fir, and called '*witches' brooms*' in Germany), or show any other forms of fungoid diseases—is certainly advisable, even at some considerable outlay, in order to prevent the propagation of the disorders, and the penetration of diseases into the stem.　But, at the same time, in coniferous forests it also diminishes the liability of the crops to be damaged by storms or by heavy accumulations of snow and ice.　And, finally, not its least utility is to be found in the fact that, in mixed forests, the judicious removal of lower branches often enables a greater density of crop to be maintained for some years longer than can possibly be the case if tending is confined solely to the operations of thinning. For species like Ash, Maple, and Sycamore, which attain maturity at a comparatively early age, this last consideration is a matter of no little sylvicultural importance.

CHAPTER X

METHODS OF STIMULATING THE INCREMENT IN TIMBER-CROPS WHEN APPROACHING MATURITY

IT is a well-known fact that, when trees growing in close canopy are thinned out, or when in the case of old crops approaching maturity a partial clearance is made, the annual increment on each individual stem naturally rises in consequence of the decrease in the number of stems drawing food-supplies and moisture from the soil, and engaging in the unavoidable struggle for light and air.

In the system of selecting only the largest trees for extraction, in standards over coppice, and in standard trees retained in high forest for a second period of rotation, such increment in growth is certainly attained. In the following cases, however, the question to be considered is mainly connected with heavy thinnings, or a partial clearance made shortly before a crop falls to the axe, in order to stimulate the remaining trees to more energetic growth in girth before the date at which the fall should take place according to the Working-Plan. It is, in fact, but the continuation of the thinnings that have been made as measures of tending throughout the whole lifetime of the crop, only it is carried out more freely on account of this being the last thinning that is to be made before the trees attain their full maturity. If less than one-fifth of the total amount of timber on the area be removed the operation may still be considered as a **heavy thinning**, whilst if one-fifth or more of the crop be thus prematurely utilized it must certainly be considered **a partial clearance.**

That such measures undoubtedly do lead to a stimulation of the annual increment is a fact not only abundantly shown by practical experience, but also proved more than half a century ago by the more eminent sylviculturists of Germany, C. Heyer, Th. Hartig, Nördlinger, and others. These authors showed conclusively that a stimulation of the increment on individual stems took place, as the natural result of all such thinnings as exceeded in degree the natural process of that unavoidable struggle for existence in which the stronger first overtops and finally suppresses the weaker individual stems, whenever such free thinnings were made in forests growing in close canopy, and at an age not exceeding to any considerable extent the normal periods of rotation under which the species of tree in question was usually grown as a timber crop. Every now and again, however, some champion steps forward to dispute what have long been accepted as facts; but the plausible deductions drawn from his observations are always, on closer investigation, found to have some flaw. Thus, for example, he may have confined his attention to the bole only, or have neglected to discriminate between the direct effects of the freer exposure to light and air, and the adventitious circumstances under which the latter may have taken place, although these concrete conditions are at times of such influence as to diminish, or even to counteract, the intended results of the thinning or partial clearance.

There is generally some very easily determinable reason when this latter measure fails to produce more energetic increment. The thinning or partial clearance may, for example, have been carried so far as actually to have interfered with the normal functions of the root-systems and the crowns of foliage of the individual trees left forming the crop. These *nouveaux riches* may often require some little time to settle down and accustom themselves to their altered circumstances; and the first form in which this makes itself apparent is very frequently rather in the energetic formation of new roots and of thicker foliage, which ultimately exert their due influence on the increment of the stem, than in the immediate thickening of the annual rings, or the immediate formation of woody-fibrous tissue. And, wherever the other main factors exerting influence on the productive capacity of the soil are not at the same time tended and well provided for, stimulated increment cannot always be expected as the natural result of diminishing the number of trees forming the crop. Thus, for example, care must be taken to prevent deterioration of the soil (where necessary, even by underplanting) in consequence of insolation, and of the drying and exhausting effects of winds, or with regard to the soil-moisture and to the soil-covering of dead foliage requisite for the formation of humus or vegetable mould.

In one case, however, the current increment of the trees may be directly decreased at first in place of stimulated. This is, when the fuller exposure to light and air leads in the first instance to abnormal increase in the production of seed, in consequence of a tendency to the formation of albuminoid substances in place of carbo-hydrates [1], as is especially liable to take place in the case of trees that were already predominating throughout the canopy at the time of the partial clearance being made. And, of course, if the partial clearance be carried too far, the increment on the remaining crop will probably be injuriously affected by the natural consequences resulting from over-exposure of the soil to insolation and the action of winds, and from exposure of the stems themselves to sunburn or scorching of the bole in smooth-barked species, to increased danger from windfall and insects, as well as to the greater damage apt to be done during the felling of the stems that are being removed, and the grubbing up of their stumps.

[1] Rinicker, *Der Zuwachsgang in Fichten- und Buchenbeständen*, 1886, p. 30.

Theodor Hartig was the first to assert[1] that, as the direct and immediate consequence of the freer exposure to light and air, without reference to the species of tree, stimulated increment on each individual tree invariably took place, though subject to the influence of the other factors determining the rate of growth. When any apparent exceptions to this rule are met with, they are neither ascribable to differences in the species of trees, nor to differences in respect of soil and situation, but are solely due to one or more of the above-indicated causes. Species of timber, soil, and situation certainly exert their influence as regards the extent to which increment may take place ; but they are not of themselves the direct or primary cause. There is, however, good reason for believing that deciduous trees are enabled by nature to avail themselves of the direct and immediate increment to a somewhat greater extent than those coniferous trees which retain their foliage throughout the whole year.

The enhanced increment need not assume the form of broader annual zones of woody-fibrous tissue along the bole. It may, and very often does, at first take the shape of considerable changes throughout the root-system and the crown of foliage, primarily and undoubtedly due to the freer exposure to light, air, and warmth. This paves the way for the succeeding form of enhanced increment, due to these changes in, and increment of, the assimilative organs, which makes itself more readily distinguishable throughout the stem and branches.

The extent to which the annual increment thus becomes enhanced varies in any particular species of tree according to the individual stem, its age and reproductive capacity, the soil, situation, and exposure on which the crop grows, the density of the canopy throughout the crop, and the development of the crown of foliage borne by the individual stem. The younger and sturdier the tree, the better developed its crown,

---

[1] *Lehrbuch für Förster*, 1861, vol. i. p. 105.

the more favourable the soil and situation may be to the thriving of the particular species of tree, and the denser or more crowded the wood, so much the more likely is the influence of the thinning or partial clearance to be readily observable. That the extent to which it is perceptible varies in the different genera of our forest trees has been above indicated and really requires no explanation. Thus, for example, Oak, Pine, and Larch, that have been grown in pure forests, in which there is a strong tendency for the trees to thin themselves naturally to such an extent as to make the canopy loose, if not broken, and to allow of each individual stem forming a larger crown than is usual in the case of shade-bearing genera like Beech, Spruce, and Silver Fir, can hardly be expected to derive so much benefit as these latter from any artificial diminution of the number of stems per acre. In many such cases, indeed, this natural thinning, especially when the Oak, Larch, and Pine are approaching the time of maturity of the crop, may have proceeded so far that any further artificial clearance might lead to diminution instead of enhancement of the increment of the remaining stems in consequence of deterioration of the soil, unless underplanting take place at the same time in order to improve, or at any rate protect, its productive capacity. As it is well expressed by Gayer [1], the founder of the modern school of scientific sylviculture—

'Whenever the partial clearance is likely to lead to interruption of the canopy of the crop, it should only be carrried out when the productive capacity of the soil is such as promises, in all its essential factors, to supply continuously the increased demands made in consequence of greater energy in the crown of foliage (transpiration and assimilation), i.e. that the soil in question is fertile, or that care may be taken to stimulate the productive capacity of the soil in some suitable way at the proper time. This can only take place through the maintenance of a good layer of humus and the careful retention of soil-moisture, and consequently in many cases only by means of *underplanting* in order to protect the soil against sun and wind.'

[1] *Waldbau*, 3rd edit. 1889, p. 571.

But in the case of the shade-bearing genera, Beech, Spruce, and Silver Fir, which (except in cases of accident) remain in dense canopy throughout their whole period of growth, right up to their normal maturity, the effects of the freer exposure to light, air, and warmth procured for the individual stems by heavy thinning, or by partial clearance when they approach the prescribed time of fall, are often during the first year marked by a two-fold to ten-fold increase in the breadth of the annual zone of woody-fibrous tissue[1].   This stimulation of increment can also in the case of these ·trees be attained with a far less degree of clearance; hence, with proper care, there need be small danger of the soil being in any way injuriously affected owing to the partial and temporary interruption of the canopy to a slight degree.   This fact is easily explained by the smaller absolute measure of light and air available for these shade-bearing species when grown in canopy of normal density.

Thus, whilst the laws regulating the increment in all species of forest trees are substantially constant, the extent to which the enhancement of increment may take place after partial clearance has, so far as observations have yet been recorded, been found to be practically in the inverse ratio to the requirements of any particular species with regard to light and freedom of crown.

The Causes of the Enhanced Increment explained.

Various causes have been assigned to the effects produced in enhanced increment after partial clearance[2].   Th. Hartig considered it to be due rather to the utilization of reserves of productive matter collected and stored up in the stem whilst it stood in close canopy, than to any increased assimilation in direct consequence of the increase in foliage that takes place when the individual tree obtains a larger growing-space;

---

[1] Grasmann, *Beitrag zur Lehre vom Lichtungszuwachs*, 1890, p. 9.
[2] Idem p. 4.

whilst Nördlinger, on the other hand, ascribed the enhancement in increment to the increase in the foliage, but laid particular stress upon what he considered the fact, that the enhancement was not so much the direct result of the increase in the mass of foliage as of the stimulated assimilative activity of the leaves and needles, with simultaneous temporary increase in the productive capacity of the soil.

The correct explanation as to the enhanced increment is probably to be found in a combination of these views. It seems much more than probable that the utilization of the reserves of productive matter (principally starch in various forms) is most likely greatly favoured by the more active assimilation of nutrients whenever the increase in the foliage takes place; and there is no reason why this should not be directly connected with the formation of the fuller crown of foliage immediately after the partial clearance has been made. It is still an open question how such reserves of nutrients are formed, how they circulate throughout the tree, and how they are finally utilized, although their existence in the parenchym-cells of woody-fibrous plants is just as well known as that of similar reserves in orchids, perennial tuberous plants, &c. The recent destruction of forests in Bavaria by the Nun moth has given Professor R. Hartig of Munich special opportunities for studying a part of this subject, particularly with regard to the Spruce [1]; and he has conclusively proved that this species of tree has much fewer reserve supplies of nutrients than the Scots Pine. In general, as is well known, such reserve supplies are far more plentiful in the broad-leaved deciduous species of trees than in conifers, and larger in the deciduous Larch than in the evergreen conifers. All the changes that take place in trees prior to the commencement of the assimilative activity in spring are ascribable to these nutrient reserves—the flowering

---

[1] See *Forstlich-naturwissenschaftliche Zeitschrift* for January, February, and March 1892.

of Hazel, Alder, and Willow, the swelling of the buds and the flushing of the young leaves, the formation of rootlets, &c. In fact, all the similar phenomena previous to the development of independent cambial activity must be ascribed to these reserve supplies of starchy and nitrogenous nutrients. The supply of reserves is usually sufficient to maintain vegetation for one complete year, and often for longer; indeed, in many cases it amounts to 7 or 8 % of the total weight of thirty-year-old timber crops. Diminution of light and foliage, consequent on limitation of growing-space, must interfere with the utilization of these reserve nutrients ; hence it is quite logical and reasonable to expect that with more growing-space, and a consequent increase of foliage, these reserves should be largely drawn upon and utilized.

That any stimulation of the productive capacity of the soil takes place in consequence of the freer penetration of the atmosphere, and the stimulus given to the formation of humus or vegetable mould, is doubtful, or even more than doubtful ; for, unless the canopy still remains comparatively dense, the soil soon becomes covered with forest weeds, which often really consume the mould, and tend to exhaust the soil, in place of allowing the timber crop to have the full benefit of the humus.

The true physiological cause of the enhancement in increment, after heavy thinnings, or partial clearance of crops of trees approaching maturity, is to be found in the stimulus to the formation of starchy matter afforded by insolation ; for the degree of activity of the assimilative organs is dependent on the purity or brilliancy of the rays of light, as the decomposition of the carbonic acid in the atmosphere can only take place when the waves of light attain a certain length. Whilst a crop is growing in full canopy, the assimilative activity of the foliage forming the lower portion of the crown is extremely small, owing to the low quality of the diffused light which is alone available ; but, when the canopy is opened up to a sufficient extent, these portions of the lower foliage, which have been practically inactive, or may perhaps even have been existing on the work done by the upper leaves or needles, reassume their normal assimilative functions and help to enhance the general increment of the tree. And, besides this increased assimilative power of the foliage, the consequent increased formation of new leaves and needles is undoubtedly of enormous influence in maintaining and increasing the general enhancement of increment throughout the stem. As remarked by König[1], the direct result of the freer exposure to light and warmth leads to the better and more energetic development of the leading-shoots, the strengthening of twigs, and the formation of twigs from buds that would otherwise most probably have remained dormant.

The enhanced increment continues until the crop once more forms close canopy. But, if it be again stimulated by a repetition of the thinning out or partial clearance, this may—if not accompanied by underplanting—be carried so far as to involve deterioration of the soil to such an extent that the beneficial influence of light and warmth on the crown is cancelled by the diminished activity of the root-system, consequently involving decreased supplies of moisture and of mineral nutrients for conveyance to the assimilative organs.

The simultaneous underplanting of crops subjected to this system of partial clearance, and especially of those of light-demanding kinds of trees—Oak, Ash, Maple, Larch, and Scots Pine—has an undoubtedly stimulating result in effecting and maintaining the enhancement of increment after partial clearance, and is of particular interest from the financial point of view. The influences exerted by the soil, the situation, and the age of the crop, exhibit themselves rather with regard to the extent of the enhancement in increment than to its production and continuance. Unless the individual trees are

---

[1] *Ueber Lichtungszuwachs, insbesondere der Buche*, 1886, p. 7.

already far past their normal maturity, and have begun to show signs of the loss of energy in growth due to a foreshadowing of the approach of senile decay, they must receive a fillip from admission to larger growing-space and freer exposure to light and warmth.   When, however, the vital energies of the trees are weakened by age or disease, normal enhancement of increment need not be looked for.   The circulation of the sap throughout the stem must already be to a greater or less extent interfered with; whilst the root-system is also unfitted to discharge the extra duties it is called upon to perform.   Trees, that are not very far past the period at which their current annual increment has culminated, as well as such as have not yet attained the culminating point, are, however, unquestionably stimulated to livelier energy, which manifests itself in enhanced increment.   But, in practice, this system of partial clearance, for the speedy formation of the more valuable assortments of timber, is seldom applicable to very young woods, or to crops that are already much older than the usual periods of rotation.   It generally finds proper scope only in mature crops of high forest and in those approaching maturity.   And in these cases the age of the crop has little influence on the length of time throughout which the enhancement of increment is maintained; for this is determined rather by the length of time which it takes the crop to form close canopy once more.   The nature of the development of the crown is, however, of greater importance than the age of the crop.   If the previous thinnings have been neglected, and if the canopy has been allowed to remain so dense that the woods are crowded, it often happens that, on receiving increased growing-space, the trees with their weakly crowns are unable to avail themselves of the advantages thereby offered, and exhibit, as may so often be seen in the case of Oaks, a tendency to the development of dormant buds, which leads to '*stag-headedness*' or death of the crown, and malformation of the bole.

Practical Advantages of the Method of Partial Clearance.

From the practical sylvicultural and the financial points of view such heavy thinnings or partial clearances should not be made until after the crop in question has completed its chief growth in height.

Along with the other advantages of natural reproduction under parent standards, the rapid increment in timber, as well as in the technical and financial value of the latter, which takes place after partial clearances made for the purpose of regeneration, has drawn considerable attention to this method as a system of treatment of woods, even when their artificial reproduction—as in the case of Oak, Maple, Larch—may generally be considered more advisable than natural regeneration. The favourable returns received from the gradual clearance of parent standards in the case of natural reproduction of Beech, and of Silver Fir in particular, as well as the financial advantages that accrue ultimately from the partial clearance and underplanting of indifferent crops of Oak, Larch, and Scots Pine when they approach maturity, have to a great extent dispelled the prejudices that existed against the system in the minds of some as being entirely inconsistent with the natural course of things, and have won for it recognition as a method of treatment worthy of adoption wherever circumstances admit of its practice. There seems little doubt that in the near future it will be carried out to a much greater extent than at present, as it combines most of the advantages of standards over coppice with decided sylvicultural and financial advantages of its own. Thus, for example, by means of this method of partial clearance, the larger and more valuable assortment of stems can be produced in a shorter time, i.e. at a less cost; whilst, instead of a high forest of normal density composed of nearly mature trees, with an average annual increment long past their point of culmination and gradually sinking, an equal and often a greater annual increment per acre is attainable after the partial clearance.   This pays as well, and often better, than if the full crop had been allowed to attain maturity, besides yielding substantial returns from the timber prematurely utilized.   When the gradual clearance of the parent standards is effected during natural reproduction, it is in no way inconsistent with what has been above said, if, before the formation of the young crop, the average annual increment of the standards is below that of normal high forest reproduced by total clearance and sowing or planting ; because, counting from the time the young crop may be considered formed, the enhanced increment on the more or less isolated standards is undoubtedly more favourable financially than when a total clearance is immediately followed by artificial reproduction.

That portion of the fixed capital which is represented by the growing crop is considerably reduced by means of the partial clearance ; and the money thus derived as intermediate returns can be otherwise utilized in order to yield interest for itself, whilst the material left standing increases in the percentage it affords.   Thus, if the partial clearance amounts to one-fifth, or one-third, or one-half of the total crop in close canopy, it has been satisfactorily proved by several authorities[1] that the enhanced increment on the area partially cleared at least equals, and often exceeds, the total previous current annual increment, notwithstanding the diminution in the number of trees thereafter forming the crop.

The financial advantages derivable are thus summarized by Grasmann[2] :—

1. The larger and more valuable assortments of timber can be produced in shorter time, and therefore more cheaply, by this method.

---

[1] Wagener, *Der Waldbau und seine Fortbildung*, 1884, p. 208; Kraft, *Beiträge zur forstlichen Zuwachsrechnung*, 1885, pp. 99 et seq.; and others.
[2] *Op. cit.*, p. 11.

2. Instead of the full crop, with small annual increment on a larger number of individual stems, three factors are favourably introduced and combined, viz. :—

 i. Good intermediate returns are available, capable of producing interest for themselves.

 ii. The enhanced annual increment on the remaining crop equals, and often exceeds, that produced by the full crop previous to its partial clearance.

 iii. The young crop produced naturally or artificially under the parent standards often practically equals in increment a young crop formed artificially by sowing or planting.

What Loudon says[1] of the Oak may be applied to all our forest trees grown as crops with a view to remunerative returns from the capital invested :—

 '*The age at which Oak timber ought to be felled, with a view to profit,* must depend on the soil and climate in which the tree is grown, as well as on other circumstances. Whenever the tree has arrived at that period of its growth, that the annual increase does not amount in value to the marketable interest of the money which, at the time, the tree would produce if cut down, then it would appear more profitable to cut it down than to let it stand.'

Whilst this is perfectly correct for each individual tree grown in the comparatively isolated positions that timber trees usually occupied in Britain at the time the above was written, it in no way precludes the possibility, which in fact we know from actual experience to be the case, that when the current annual increment per acre has sunk below the point up to which it is not unprofitable to allow a crop to remain in the full normal canopy of its species, it can be stimulated and rendered more profitable, often very considerably so, by the above-indicated method of partial clearance with a view to the earlier production of valuable stems of large girth. In regular high forest, with natural reproduction, the time of regeneration

---

[1] *Arboretum et Fruticetum Britannicum*, 1838, vol. iii. p. 1809.

is often, in order to attain good boles of valuable dimensions, prolonged till after the period when energy for the formation of seed, or the natural reproductive capacity is already somewhat weakened. By means of partial clearances, however, the regeneration can be favoured during the chief natural seed-producing period of growth ; whilst, at the same time, as good dimensions of timber may be obtainable as in the regular high forest, and in a more profitable manner. Thus, for example, in the reproduction of Spruce in sheltered localities—where windfall is not likely to occur, and where the period of reproduction may be taken to be usually about a hundred years—if partial clearance be made about the seventieth year, during the time of the most active production of seed, the financial position of soil plus crop during the succeeding thirty years will almost, beyond question or doubt, be much more favourable than if the crop had been allowed to stand as regular high forest till the full period of rotation originally fixed in the Working Plan had been attained; for it is almost certain that the partially cleared crop will contain a relatively higher number of boles of valuable assortments than would have developed themselves under the system of regular high forest. And what is said here with regard to Spruce is still more applicable in respect to the other two great shade-bearing genera, the Silver Fir and the Beech. The whole system is, in fact, merely a modification of, and improvement on, the older system of natural reproduction under parent standards. But, whilst in the latter case the seed fellings (partial clearances) before and during the time of seeding, and the gradual clearances after the formation of the young crop, took place mainly with the object of effecting natural regeneration, the partial clearances under the new method are made with the distinctly avowed intention of stimulating the standard crop to enhancement of the annual increment, and to the yield of more favourable returns from the capital represented by soil and growing crop. In the former case formation of a young crop was the motive, and enhanced increment on the parent standards the effect; in the latter, enhancement of annual increment is the object, and simultaneous natural reproduction of shade-bearing kinds of trees is the result of this sylvicultural operation.

Rules for the Conduct of Partial Clearances.

With the light-demanding species of trees, when they are not grown in mixed forests along with shade-bearing species, underplanting is almost a necessity after heavy thinnings or partial clearances, in order to maintain the productive capacity of the soil. The manner in which this method of treatment is applied to the various chief genera of our light-demanding forest trees—Oak, Pine, and Larch—is as follows :—

Oak.—When pure forests of Oak have passed through the regular processes of thinning, and approach the time for a partial clearance taking place, their canopy is in general somewhat light and thin, although perhaps not broken; and the undergrowth is usually already beginning to form canopy for itself below the older crop. To avoid the formation of twigs and shoots from the dormant buds along the stem, often leading to '*stag-headedness*[1],' the partial clearance should not be made by one fall, but is better attained by one or two heavy thinnings conducted to the extent deemed advisable for each individual crop. The time at which it should take place is also dependent on the special circumstances of each crop, special regard being given to the time of underplanting and the development of the undergrowth, the energy of growth of the trees, and the nature of the soil and situation. As, however, the efficacy of this method of treatment is all the more apparent the earlier

[1] In his *Lehrbuch der Baumkrankheiten*, 1889, p. 14, R. Hartig remarks that :—' Oaks, which have grown up in close canopy along with Beech and have only a slightly-developed crown of foliage, acquire a predisposition towards a drying-up of the top of the crown (*stag-headedness*) when they become fully exposed to light and air; whilst, under similar conditions, trees with well-developed crowns do not suffer from this disease.'

it can be begun, the thinnings should, if possible, be commenced about the fortieth to sixtieth year. They should at first
be confined to the young stems of backward growth, and to
such as are never likely to attain good marketable shape. The
oftener the thinnings can be repeated at short intervals the
better, as any sudden exposure of the boles should be carefully
avoided.

About ten or fifteen years after these preliminary operations
have been begun, the main partial clearance will in most cases
seem advisable; for by that time the crop should be in a state
of active increment, due to the favourable influences of the
undergrowth and the previous thinnings. Here, again, as also
whenever further partial clearances are considered necessary
later on, the fall is in the first instance confined to trees that
have been damaged by organic or inorganic agencies, or that
do not continue to yield satisfactory increment; and not until
all such unremunerative material has been removed should
any sound trees in energetic growth be cut out.

When the main partial clearance has taken place about the
seventieth year, and has been followed by minor gradual
clearances at intervals of five years at first, then of ten, and
later on of fifteen, the preliminary yield or intermediate returns
thus obtained should amount, according to Kraft[1], respectively
to about 288 to 628 cubic feet, then to 720 to 1,080 cubic feet,
and ultimately to 1,800 cubic ft. per acre, inclusive of branchwood. Past experience seems, according to Gayer[2], to justify
the hope that by this method of treatment the same dimensions of bole can be attained in about 120 years as are
obtainable in pure forests worked with a rotation of 200 to
240 years, provided that up till about the hundredth year the
rate of increment has been maintained at 3 to $3\frac{1}{2}$ %, and after
that at 2 to $2\frac{1}{2}$ %, conditions which are quite conformable with
the actual results obtained on suitable soils. The crop of Oak

---

[1] *Aus dem Walde*, vol ix., p. 80.
[2] *Op. cit.*, p. 574.

then ultimately harvested numbers about thirty-six to forty-eight stems per acre, which on good soil yield valuable and remunerative assortments of timber[1]. Where soil and situation are good enough to fulfil the demands made on them, there can be little doubt that this system of underplanting with a shade-bearing species, and partially clearing the main crop from time to time, is the method best calculated to produce timber of the highest value for technical purposes, and to yield the most favourable returns financially from the capital represented by the soil, together with the growing crop.

Scots Pine.—For the practice of this method a good deep soil is necessary, as it differs essentially from the custom, common on poorer qualities of soil, of thinning out crops during their twentieth to fiftieth year of age and underplanting them with Spruce or other suitable shade-bearing species. In the latter case they are often allowed to grow on to maturity without any further thinning or interruption of the canopy to speak of; but, though they continue to improve in growth and form, this is due to the beneficial influence of the undergrowth both on the soil and the standards. As a good formation of the crown is requisite for Scots Pine, in order to enable it to continue for a long time in active energetic increment, thinnings should be made towards the end of the pole-forest stage of growth, and afterwards repeated whenever there seems any danger of the crowns interfering with each other. Underplanting should also be here carried out as soon as the thinnings have taken place with a view to the formation of

---

[1] The following passage from Loudon's work already quoted (p. 1809) is interesting, although not quite intelligible as regards *the largest scantlings* being produced at 130 years of age—though if *larger* be read in place of *the largest* the statement readily becomes intelligible :—

'A writer in the *Gardener's Magazine* states that Mr. Larkin, an eminent purveyor of timber for ship-building, stated, when examined before the East India Shipping Committee, that, in situations the most favourable for ship timber (the Weald of Kent, for example), the most profitable time to cut Oak was at ninety years old; as, though the largest scantlings were produced at 130 years' growth, the increase in the forty additional years did not pay 2 per cent. (*Gard. Mag.* vol. xi. p. 690.)'

good individual crowns. By this means, in place of having a pure forest formed, according to Weise's Yield-Tables[1], on soil of average quality at eighty years of age by 317 stems per acre (having a mean average girth of 29 inches at breast-height, which would only be attained by about 40% or 127 stems, with a form factor of 0.45 and a current annual increment of only 44 cubic ft. per acre), it can hardly be doubted that a smaller number of trees with enhanced increment and the finer dimensions of stem effected by the comparatively rapid thickening of the bole at the top end must yield more favourable financial returns on the capital invested in timber production. Reliable comparative data are, however, unfortunately not yet available to illustrate this point.

Larch.—In regard to general treatment by partial clearance with a view to the speedy production of large-girthed timber, the Larch has much in common with the Oak ; but the underplanting may sometimes take place more advantageously with the Silver Fir than the Beech. So far as experience goes, this method of treatment is well adapted to the cultivation of the Larch on good, deep soil ; and it is most satisfactory when the partial clearance and underplanting takes place as early as the twenty-fifth to thirtieth year. When the previous tending of the plantations has been good, and the thinnings have been regularly made, the main partial clearance can be heavier than in the case of the Oak, so that subsequent clearances are often almost unnecessary. On good soil from sixty to seventy trees per acre may remain to form the final yield of the standard crop ; and wherever soil and situation are at all suitable for the growth of Larch, sixty- or seventy-year-old crops can often yield an increment of 3 to 4% up to that age, when they should already show very good marketable dimensions.

As the **Beech** is not in favour or demand as a timber-tree of large dimensions, this method is only apt to find any practical

---

[1] *Yield-Tables for the Scotch Pine*, translated by W. Schlich, Ph.D., 1888, pp. 18, 19.

application with the Spruce and the Silver Fir among shade-bearing genera ; and as they show many points of similarity, these two trees may be considered together.   Where, however, there is any likelihood of windfall or damage from storms, this method is totally unsuited for the Spruce.

Spruce and Silver Fir.—The way towards partial clearance should be paved by preliminary operations carried out about the twenty-fifth to thirtieth year in such a manner as to let the predominating poles have a clear space of 2 or $2\frac{1}{2}$ ft. around the crowns.   These favoured individuals should stand about 15 to 17 ft. apart, all the rest of the crop being thinned out in the ordinary manner customary in pure forests without interruption of canopy.   During the thirtieth to fiftieth year, when the influence of this measure has made itself apparent in the re-formation of close canopy, the main partial clearance takes place in a somewhat similar manner.   The stems between the favoured individuals should be freely thinned so as to permit of the formation of an undergrowth either naturally, or where necessary artificially, wherever the minor portion of the crop is of itself insufficient to protect the productive capacity of the soil.   When the favoured stems have at breast-height a girth of about 36 to 40 inches, which they should attain between the sixtieth to eightieth year, the marketable trees can then be gradually cleared away on the same principle.

Another method, practised in the neighbourhood of Salzburg in Western Austria throughout mixed forests consisting of Spruce, Silver Fir, and Beech, consists in carrying out the thinnings once every ten years after the crop has attained thirty years of age, and in making them heavier each time the operation is repeated, until the main partial clearance is made about the sixtieth or seventieth year, leaving 120 to 160 stems per acre, all of them conifers and in energetic growth.   Under such a crop natural reproduction can easily be effected, so that about twenty years later the standard trees are of large marketable dimensions, whilst the young growth may vary from

5 to 15 ft. in height.   On the removal of the standards the
blanks thus formed in the new crop can easily be filled up
artificially, and may be advantageously utilized for the formation of mixed forests by the introduction of other species.

Actual measurements of crops thus treated are not yet
available for comparison with the yield-tables published for
pure forests; but the approximate data already given may be
contrasted favourably with the more accurate information contained in these latter.   According to v. Baur, pure forests of
Spruce of 120 years in age and growing on good soil have
288 stems per acre, the mean girth of which at breast-height is
39 inches, but this is only attained or exceeded by about 40 %
or 115 stems; the mean average stems have also been found
to have a form-factor of 0·48, which is undoubtedly lower than
that of stems treated by the method of partial clearance.
Again, according to Lorey, the Silver Fir, when grown in pure
forest on good soil, yields 200 stems per acre with a mean
girth of about 52 inches at breast-height, which also is only
attained by about 40 % or eighty stems per acre.

As has already above been pointed out, this method of
partial clearance with the distinct object of stimulating the
remaining crop to the speedy development of the larger and
more valuable assortments of timber cannot be successfully
attempted with regard to any species of tree when once its
natural energy of growth has practically abated to any considerable extent, or on the poorer classes of soil; but wherever
the trees are still capable of stimulation in respect to increment, and the productive capacity of the soil can be easily
safeguarded against deterioration, the method has decided
practical advantages which should strongly recommend themselves to the woodland proprietors of Britain, where sylvicultural operations on any extensive scale should only be
conducted on strictly financial principles.

CHAPTER XI

THE PRACTICAL EFFECTS OF UNDERPLANTING

THE conservation of the productive capacity of the soil is one of the first fundamental principles of Sylviculture; for the sustained yield of the best classes of timber can only reasonably be expected when measures are taken to safeguard all factors determining the quantity or quality of the food-supplies available for the root-systems. One of the first practical steps in this direction consists in the protection of the soil against insolation, and against loss of soil-moisture, by means of the maintenance of a good, close leaf-canopy and the consequent thicker layer of dead leaves. This both forms humus or mould for the benefit of the soil, and also acts mechanically to prevent evaporation from drying winds. Beech, Spruce, Douglas and Silver Firs, and Black Pines—species of trees that are densely foliaged and therefore naturally capable of bearing shade well— are endowed with good soil-protective qualities throughout the whole period of their development until they attain their technical and financial maturity; but most of our other valuable forest trees, Oak, Ash, Maple, Birch, Larch, and Scots Pine, lose their power of conserving the productive capacity of the soil long before they attain their maturity, owing to the natural demands they exhibit for increased growing-space and relatively greater coronal development when once they have nearly completed their main growth in height. These latter species, in fact, prove themselves light-demanding in place of shade-bearing; and, if measures are not taken to counteract the consequences of the gradual interruption which takes place in the leaf-canopy, the effects of insolation and of exposure to the freer action of wind and rain on the soil, tend to diminish its productive power, and must ultimately lead to its deterioration.

To obviate the consequences of this natural tendency on the part of the lightly-foliaged species, which is not reconcilable with the conservation of the productive energy of the soil, the best plan capable of adoption is to underplant them with shade-bearing species for the purpose of maintaining the soil cool and moist, and of improving it through the larger quantities of mould formed by their dead foliage.

In most of the open forests of Oak, Larch, or Pine, for example, there will usually be some sort of natural undergrowth found under the comparatively light shade of these forest trees; but the practical influence exerted by an underwood of brushwood, scrub, or berries, is by no means so beneficial as that of undergrowth formed either naturally or artificially by Beech, Hornbeam, Sweet Chestnut, or Sycamore—or by evergreens like Spruce, and Silver, Douglas, or Nordmann's Firs[1].

The first occasion on which underplanting is known to have been carried out under Oak, with the express intention of

'giving a covering of dead foliage to a soil already overgrown with whortleberry, and thus strengthening its productive capacity, and stimulating the diminished increment of the Oaks,'

took place in the Spessart forest in central Germany about

---

[1] Under Firs are included only the genera *Picea*, *Abies*, *Tsuga*, and *Pseudotsuga*, including Spruce, Silver Fir, Hemlock, Nordmann's, Menzies, and Douglas Firs, but not the Pines (*Pinus*) incorrectly though perhaps more commonly called Firs, e. g. Scots Fir for Scots Pine. Etymologically, *Fir* is properly the name of the Scots Pine, from the old Anglo-Saxon *furh*; but if we retain it, then we must coin distinctive names for Spruce and Silver Fir, which are distinctly different genera not indigenous to Britain, and only introduced during the seventeenth century. For these trees we have never adopted any generic names such as are current in France and Germany, whence they were brought over.

1840[1], since which time the practical and financial advantages of the measure have been duly observed and recognized.

Copse or coppice under standards has, of course, for centuries been cultivated both in Britain and on the Continent; but in 1840 the first case of *underplanting* took place solely for the benefit of the standard trees, and not for financial considerations mainly dealing with quick and frequent returns from the coppice underwood.

There are, however, two methods of effecting the end in view; for the underwood may be formed either for soil-protection only, or else with the intention of allowing it to grow up into tree-forest to be ultimately felled along with the present stock of standard trees, in which latter case the crop really attains maturity as a mixed wood, and therefore participates in the advantages offered by such over pure forests[2].

As on most matters, whether practical or scientific, concerned with Forestry in Germany, a paper war has been waged on this subject. Whereas the benefits of underplanting should have been patent to all, there were not wanting some (headed by Professor Borggreve) who contended that, in place of stimulating the standards to greater increment, the practical effect of the underwood was to consume a large portion of the food-supplies available in the upper portion of the soil, whilst its formation often necessitated a considerable outlay[3]. Owing to these differences in opinion, the matter was taken up for careful investigation by the experimental section of the Forest branch of the Faculty of National Economy in Munich University.

The opponents of underplanting based their objections on three different classes of experiments:—

1. Experiments made with standards in copse (i. e. over coppice).

---

[1] See *Monatschrift für Forst- und Jagdwesen*, 1874, p. 1.
[2] See Chapter VI. *On the Advantages of Mixed Woods over Pure Forests.*
[3] *Forstliche Blätter*, 1877, p. 20; 1883, p. 41.

2. Observations as to the rate of growth in crops that had been underplanted for some time, and a comparison of that with the rate of growth in other portions not underplanted.

3. Clearance of underwood in crops underplanted, and comparison of the next year's increment with that on those under which the underwood had been left untouched.

So far as the first class of experiments was concerned, Borggreve found that the stems exhibited increment in the breadth of the annual rings only for five years, whilst a decrease set in whenever the underwood began to close up and form canopy, and that the decrease became greater as the canopy grew more dense. He therefore concluded that the initial enhancement was solely due to the more complete insolation of the trees and the diminution in the number of plants drawing their food-supplies from the soil, and denied that it could be due to any better and more rapid formation of mould by the dead foliage covering the soil. But he over-looked two points—*firstly*, that the degree of insolation of the crowns of the *standards* could in no way be affected by the removal of the *coppice*; and *secondly*, that the demands on food-supplies from the soil made by the coppice during the first few years of its rapid growth would certainly be *at least* equal to (and most probably higher than) were made later on when it was growing in close canopy. When these points are duly considered, it seems evident that the cause of incre-ment must certainly be assigned to the advantageous formation of humus or mould when the dead foliage was exposed to the decomposing influences of sun, rain, warmth, and fungoid growth. He maintained—with a show of reason, but incor-rectly—that these processes were retarded rather than favoured by the clearance of the coppice. Full, clear consideration of the actual facts warrant the opinion that the increment took place on the standards owing to the larger formation of humus from the dead foliage which could not decompose normally under the dense shade of the coppice, and that the subsequent decrease, five years later, was due to the fact that the richer supplies of nutrient salts, rendered soluble and available as food in the soil owing to the larger quantities of humus formed after the fall of the coppice, began to get exhausted after about four to five years' time.

The argument is also, however, open to another grave objection; for the observations were confined to the lower end of the stem, and were not continued all the way up the bole. Now, Weise has shown that on Oak standards increment becomes stimulated near the base of the stem, but that, when undergrowth gets over twenty years in age, the basal increment decreases, whilst that near the top of the bole becomes enhanced—or, in other words, the top-girth becomes relatively larger, so that the stem tapers less and has a higher technical and financial value.

The data with which the deductions under the second class of experiments were supported, were found on examination not to be reliable, as they were taken from mixed crops of Pine and Spruce of equal age (probably formed by sowing), in which the struggle between them had been so keen as to prejudice the development of both species. Hence, in the Scots Pine wood classed as *without underwood*, the Spruce had been removed in the thinnings; whilst, in that shown as *with underwood*, the Spruces were not truly underwood, but were suppressed poles still struggling for life with the Pines, which therefore showed less increment than they would have exhibited under normal conditions. In other data, the areas reported on were so small that general deductions were not justifiable; moreover, the average height of the stems (which in general affords a very sound means of judging of the quality of the soil [1]) showed that the areas not underplanted were somewhat better than the others subjected to consideration.

With regard to the third class of experiments, Michaelis

---

[1] Grebe, *Ertrags- und Betriebsregelung*, 2nd edit. 1879, p. 114.

found[1] that in Oak forest, with an eighteen-year-old under-growth, mostly of Hornbeam and Hazel in close canopy, the increment was practically the same as was to be found on the trees in another part of the same crop, which had been fenced off as a game-park, and had been browsed on by deer to such an extent that there really was no underwood to speak of. As accurate details were not collected with respect to the nature and physical properties of the soil and situation (as to depth, moisture, exposure, &c.), it would not be justifiable to argue from such meagre data that undergrowth was unnecessary for the retention of the productive capacity of the soil. It is, indeed, quite true that the main food-supplies are derived from the lower layers of soil only; but, as those are only avail-able whilst dissolved in water, and as the layer of mould assists percolation by preventing the rapid off-flow of aqueous precipitations, it can easily be understood how beneficial must be the action of humus and of a good soil-covering on all such soils as are apt to suffer from want of moisture.

Among the advocates of underplanting, Frömbling in 1886 published[2] measurements made on Oaks of 160 years in age growing on old grazing land, but which had been underplanted for thirty-five years. In three stems growing close together and not thinned, the breadth of the annual rings had sunk to their minimum at the time of underplanting, but had broadened as the undergrowth formed canopy, and became broader than any of the annual zones during any of the previous fifty to sixty years when the natural energy of growth of the trees was at its maximum. Other stems, which had been thinned or partially cleared, formed considerably broader zones at first in con-sequence of the freer exposure to light; but this fell off again for three years, and then recommenced in response to the improvement in the productive capacity of the soil consequent on the formation of canopy by the underwood and of humus

---

[1] *Forstliche Blätter*, 1884, p. 345.
[2] *Zeitschrift für Forst- und Jagdwesen*, 1886, p. 632.

by its dead foliage. The deductions he drew were therefore diametrically at variance with those previously arrived at by Michaelis; whilst the data on which they are based leave no possibility of differences existing with regard to the quality and physical conditions of the soil and subsoil.

But, before then, valuable data had already been contributed by Rünnebaum[1] respecting two areas under Scots Pine, one of which (120 years) was underplanted with Beech, and the other was without underwood (110 years). The differences observable as to mineral composition were inconsiderable; but the former had a thick layer of humose soil containing nearly double the quantity of mineral nutrients that were found in the thinner layer of humose soil on the area without underwood. The yield and returns from these areas (without reckoning 343 cubic ft. of Beech sold from the former for £27 13*s.* 0*d.*) were as follows:

Age of Crop.	Outturn.	Annual average Increment.	Selling price.			Rate per cb. ft
Years.	Cub. ft.	Cub. ft.	£	*s.*	*d.*	Pence.
120	14,280	119	229	2	0	3·8
110	13,370	121½	149	2	0	2·7

The slight difference in average annual increment in favour of the younger crop (without underwood) is of course due, not to differences in the soil and situation, or to greater increment without undergrowth, but to the fact of the larger quantity of timber removed in the thinning at the time of underplanting What must strike one most in the above data is the much higher rate per cubic foot obtained from the standard Pines, grown over underwood, in consequence of the stems being straighter, smoother, less tapering, and with a proportionately larger development of the summer zone in each annual ring. Prof. R. Hartig of Munich was the first to enunciate the theory

---

[1] *Zeitschrift für Forst- und Jagdwesen*, 1885, pp. 156 et seq.

that the underwood, by keeping the soil moist and cool in early spring, delayed the early awakening of vegetation and of assimilative processes for about a fortnight at least. This prevents the commencement of growth until somewhat warmer weather has set in, and thus stimulates the energy of the assimilative organs, and leads to the formation of thicker cell-walls in the woody fibrous tissue. To be unbiassed, it must be pointed out that the comparative freedom from branches had nothing to do with the undergrowth. It was due to the fact that the crop was originally a mixed wood of Pine and Beech, from which the latter was cut out about the sixtieth to seventieth year, after the chief growth in height had already been completed, whilst a spontaneous undergrowth of Beech took possession of the soil; for it is a well-known fact that this species regenerates itself better under Scots Pine than under the denser shade of parent standards.

The investigations conducted by the Munich authorities are worthy of great respect; for they were made in a truly scientific spirit, free from bias either one way or another. They not only take into account the breadth of the annual rings at various heights above the ground, but also the superficial increment, the increment in height, and the total increment in cubic contents—without all of which reliable data are not obtainable. Where they are weak is that, although over 7,500 acres of high forest have been underplanted and might be chosen from, yet, unfortunately, areas without underwood, but absolutely similar as to soil and situation, are wanting for accurate comparison. Three sets of investigations were conducted, with the following results.

The first series was made in Oak forest (*Q. pedunculata*) of eighty-two years of age, which had been underplanted with Beech at fifty-four years of age (twenty-eight years previously). The underwood had partly grown up into the crowns of the standards, and the soil had a dense layer of dead foliage and mould. The records of the Working Plan showed that at the time of underplanting a heavy thinning or partial clearance had taken place. This was of itself at first sufficient to stimulate the trees to the production of broader annual rings, but only in such a manner as to induce a tapering form of bole owing to broader zones being formed near the base than higher up the stem. But ten years later, when the underwood began to form canopy, the breadth of the annual rings diminished considerably; and, in the case of four out of six sample stems, this was succeeded by a slight increase again during the third decade. During the second decade, whilst the increment was slightly diminished, there was a distinct tendency to approximate in mode of increment towards what obtains in close-canopied forest. That is to say, in consequence of the underwood forming canopy, the annual rings near the base of the stem became relatively narrower whilst those near the top of the bole became relatively broader, with the result that the bole became less tapering and consequently better-shaped, technically more useful, and financially more valuable. This improvement, often a very slight one only, is due to the fact that, owing to the soil, and consequently the root-system, being kept cool, the latter is not stimulated to the extraction and absorption of such large quantities of nutrients as if warmed by free insolation. Hence the quantity of assimilated nourishment and constructive matter which is conducted downwards to the base of the stem and to the root-system is less in amount than under these other circumstances; and as the total amount of food-supplies is diminished, whilst the demands of the top part of the bole for nourishment practically remain constant, less remains to be utilized for constructive purposes towards the base and throughout the root-systems of the trees.

The second series of investigations was made on two Oak standards of 105 years of age, which, after having stood for about twenty to thirty years with their crowns freely exposed to light and air, were underplanted at sixty-five years of age with Beech. The latter is now about forty years old, forms close canopy, and is already partially forcing its way into the lower portions of the crowns of the standards. The Oaks were rather short-stemmed; for not only was the soil somewhat inferior, but the trees had also been allowed to develop without forming close canopy. Records showed that the soil had formerly been overgrown with whortleberries, though at the time of the investigations it had a thick layer of dead foliage and humus. The underwood was formed by sowing during 1843 to 1847, immediately after a thinning took place, which was followed by a clearance of all diseased stems during 1853–1856.

The measurements made on the stems, when felled, showed that (1) about eight years after the formation of the undergrowth the rate of increment had sunk considerably, after having increased at the time of its formation owing to the thinning; (2) the effects of these two thinnings or partial clearances continued for about two decades to induce tapering growth with larger annual rings near the base than the upper part of the bole; (3) during the third and fourth decades after the formation of the underwood the basal increment was lower in No. 1 and the same in No. 2 as previously, whilst in both cases the increment in the upper portions of the trees was somewhat enhanced; and (4) finally, during the fourth decade, the superficial increment of the various sections, and the total increment in cubic contents, were practically about twice as great as they were throughout the decade previous to underplanting. This last result completely disposes of the argument that the growth of the standards is prejudiced by the underwood consuming a large share of the nutrients in the soil. At the same time it is hardly open to reasonable doubt that the gradual improvement in the shape of the bole is due to the underwood growing up into the crown, and interfering more or less with the activity of the foliage of its lower portion.

The third series of investigations was carried out in an Oak wood eighty-eight years old, whose past history was accurately traceable. Formed by sowing on old arable land, its growth and development were good at first; but about its fortieth year the soil became overrun with whortleberry, the increment of the crop fell considerably, and its appearance was unpromising. Without any thinning being made, beech-mast was sown out in 1840 for underwood, to which during 1844–1847 a heavy thinning or partial clearance succeeded with simultaneous underplanting (at forty-six years of age). Light thinnings were afterwards made in 1853, 1859, 1873, and finally in 1887, when about eighty stems per acre were removed, leaving about 300 per acre on the average, which had formed canopy again. The effect of the first thinning was hardly sufficient to affect the breadth of the annual rings or the general shape of the bole; but it was still traceable in the predominant and dominant stems till the third and fourth decades.

The measurements made on sample stems showed that the increment in height diminished after the heavy thinning or partial clearance, but eventually attained its former level, and in some cases exceeded this. That is to say, the energy of growth in height was diverted and utilized in the coronal expansion, whilst it was re-directed into its original channel when the crowns approached again to form close canopy.

Without giving any of the statistics and measurements recorded—which can only be exhibited, for each series of investigations, in an extensive tabular form that might be out of place here—the following are the results arrived at, as published in a pamphlet by Dr. Kast, Lecturer on Forestry in Munich University[1]. When crops of light-demanding forest trees are underplanted, and the underwood begins to form close canopy, the effect is to produce a diminution in the breadth of the annual rings near the base of the stem,—but this only affects the upper portion of the bole either to a less extent, or not at all,— whilst the total annual increment is not always decreased. If any such decrease be considerable, it is probably due to

---

[1] *Ueber den Unterbau und seine wirthschaftliche Bedeutung,* 1889.

re-formation of close canopy, and consequent sinking of the assimilative power of the lower and inner foliage.

With regard to Prof. R. Hartig's theory, that heavier and better timber is formed when underwood protects the soil, Rünnebaum's experiments confirmed its correctness, at any rate so far as Scots Pine is concerned, but without yielding other final results of a satisfactory nature when criticised from a purely academic point of view. In a general way, it might, however, be deduced that the undergrowth, though in other physical respects of favourable influence, does not (as has been hitherto generally accepted) directly stimulate to enhanced annual increment,—owing to the fact that it utilizes a portion of the soluble nutrient salts contained in the soil for its own growth and development. But the quantity of nutriment thus diverted from the standards is, especially on the better classes of soil, not sufficient to justify the objections of the opponents of underplanting. Their argument, that after heavy thinnings or partial clearance a spontaneous growth of grasses and weeds covers the ground, and annually restores to the soil the nutrients previously extracted from it, ignores the researches of Vonhausen, who proved[1] that a soil-covering of grass extracts from four to five times as much mineral food (salts) from the soil as is required by Beech underwood, whilst the humus or mould formed by such herbage is almost entirely consumed by the soil-covering of weeds annually without the standard timber trees being able to reap any benefit therefrom. It must also be remembered that the quantity of nutrients annually extracted from the soil by the underwood is comparatively so small as hardly necessary to be taken into consideration, at any rate on the better classes of soil, so far as the food-supplies of the standard trees are concerned; whilst the influence that may thus be exerted on the productive capacity of poorer classes of soil must be more than outweighed by the greater increment due to the thinnings or partial clearances in

[1] *Allgemeine Forst- und Jagd-Zeitung*, 1872, p. 1.

tracts underplanted, as compared with similar woods not treated in this manner. There are, of course, circumstances and situations in which undergrowth might make relatively higher demands on the soil than are made by the standard trees, as, for instance, any endeavour to underplant Scots Pine with Beech on poor, dry, sandy soils,—an attempt which would be diametrically opposed to the true sylvicultural characteristics of the latter kind of tree.

Of at least equal importance to the amount of annual increment on the standards, the quality of the timber produced next demands consideration. Its technical utility is, apart from such very special cases as formerly obtained when Oak-crooks were in demand for ship-building, mainly determined by (1) straightness of growth, (2) full-woodedness of the top end or high-form factor, (3) freedom from knots and branches, and consequent evenness of texture, (4) proportion of heartwood to sapwood, and (5) height of specific gravity—for with regard to any samples of wood of the same species the heaviest is *ceteris paribus* the most durable[1], as its specific gravity is dependent on thicker deposits of ligneous substance on the cell-walls. This last point is determined by the proportion of the denser summer zone to the more porous spring zone in the annual ring, the former being formed when the assimilative process is carried out most thoroughly during the hottest months of the year. And, as the practical effect of the under-wood is to delay the awakening and commencement of active vegetation for at least a fortnight, owing to its maintaining the soil and the root-systems of the standard trees cool and moist under its shade, it therefore exerts a most beneficial influence on the timber produced. When heavy thinnings or partial clearances have been made, this mechanical hindrance to insolation and to early stimulation of active vegetation is all the more important, as, from the comparative isolation of the standards, an early commencement of the transpiratory and

[1] Gayer, *Die Forstbenutzung*, 7th edit., 1888, p. 66.

assimilative functions would naturally be induced, which would certainly lead to the production of a larger quantity of timber but of lower specific gravity, and consequently of less technical utility, than was being formed previous to the artificial interruption of the leaf-canopy.

And, at the same time, the underwood, as it gradually grows up into the lower portion of the crowns, is of practical utility in cleaning the boles by interfering with the vital functions of the lower branches, and suppressing the activity of their foliage. It thereby also helps to concentrate and confine the constructive functions within the upper regions of the bole chiefly, so as to increase the girth of the top-end relatively to that of the base, and thereby enhance the technical and financial value of the stem. These advantages must of course be counterbalanced if the underwood be allowed to grow up so far as to interfere with the normal development of the standards or to cause partial suppression of the crowns.

The quantitative relation of the heartwood to the sapwood in tree-growth is still an unread chapter in the great book of the laws of nature. Up to the present, practically nothing is known about this matter. Künnebaum and others have shown that in Pine woods underplanted with Beech the proportion of heartwood to sapwood is greater than in Pine standards growing without underwood ; but as similar benefits can be proved in the case of mixed woods of Pine, Spruce, Beech, &c., there is as yet no sound basis for justifying the assumption that underplanting specially favours any shortening of the period of semi-activity which the older sapwood has to pass through before attaining its full maturity as the kern or heartwood of the tree.

Undoubtedly the greatest influences exerted by undergrowth are to be found, *firstly*, in the richer stores of humus or leaf-mould with which it provides and enriches the soil through the decomposition of its annual fall of leaves and twigs, whereby all the physical properties of ordinary woodland soils become

greatly improved, and the previously extracted nutrient salts are
again largely restored in a convenient, easily absorbed, soluble
form, and *secondly*, in the more thorough aeration of the soil
by means of the root-systems of the underwood. It is true that,
even when forming undergrowth and enjoying only comparatively limited exposure to light, air, and warmth, Beech undergrowth still transpires enormous quantities of moisture drawn
from the soil, and intercepts a high percentage of the precipitations before they reach the ground; but against this must be
duly weighed the greatly advantageous influence exerted by
the strongly hygroscopic humus in retaining the moisture
reaching the surface, and in allowing it facilities for entering
and percolating the soil, in place of running rapidly off into
the nearest water-channels, and of being quickly evaporated by
insolation and the freer play of winds. As a matter of practical
result Ramann found that, on the areas where Rünnebaum's
experiments were conducted, from the middle of May till the
end of August there was always more moisture immediately
underlying the humus in the Pine woods underplanted with
Beech than in those with no undergrowth; and that, down to
2 ft. below the surface, the soil was moister in the former for
the first half of that time but drier throughout the later half,
whilst at a depth of about 28–32 inches the soil under the
pure Pine woods was always moister [1]. His deductions therefrom were that, under the light shade of the Scots Pine, the
spontaneous growth of berries, grass and other weeds exhausted
the superficial supplies of moisture to a greater extent than the
Beech underwood, whilst the layer about 20 to 24 inches below
the surface was called upon to furnish supplies to both Beech
and Pine; whereas in pure Pine woods the lower soil received
larger quantities of moisture from above after the grasses had
completed their development than when a soil-covering of
Beech underwood existed. Wollny [2] also found, in independent

---

[1] *Zeitschrift für Forst- und Jagdwesen*, 1885, p. 172.
[2] *Forschungen auf dem Gebiete der Agricultur-Physik*, vol. x. p. 261.

experiments, that the quantity of moisture under a soil-covering of living plants is always less during the period of active vegetation than at a similar depth on naked soil (in consequence of transpiration); that the influence is quite traceable down to the lower layers; that the quantity found is less when the plants stand densely (without being inversely proportional to the density); and that a soil-covering of humus and dead foliage, up to 2 inches in thickness, favours percolation from above downwards as compared with naked soils.

As Beech, Hornbeam, and Silver, Nordmann's and Douglas Firs have all very much the same sort of heart-shaped, moderately deep root-system, the practical effects of underplanting with these different species will ordinarily be about the same; whilst differences will be noted when the shallow-rooting Spruce is selected as the underwood on soils for which it seems best suited. In the latter case the woods will still show more moisture than pure woods throughout the upper layer of soil, for the Spruce undergrowth will draw its main supplies of water from the layer from 12–18 inches below the surface; whilst, below that, the quantity of moisture will be much the same, at any rate during the first half of the annual period of vegetation. But, owing to the great density of an underwood of Spruce, a large proportion of the atmospheric precipitations will be intercepted without ever having any chance of percolating to the lower soil; hence the deeper layers will generally have less moisture than similar soils without underwood. During the autumnal and winter period of rest the soil under broad-leaved, deciduous underwood must (*ceteris paribus*) show larger supplies of moisture than under a coniferous undergrowth with persistent foliage, which not only intercepts a larger percentage of the precipitations, but also requires water from the soil for purposes of transpiration.

Dense undergrowth throughout a whole wood will of course withdraw larger supplies of moisture from the soil than that which forms only a light canopy. Hence underwood may sometimes be utilized to the greatest practical advantage if only formed here and there where there is no good layer of dead foliage or moss; for this may often give the most practical combination of protection against sun and wind and of formation of humus with least drawback on account of consumption of soil-moisture and soluble nutrient salts. And although the old idea, that soils covered with underwood were invariably moister than similar soils without undergrowth, has not proved itself to be unconditionally correct, yet the advantages of the undoubtedly better layer of dead foliage and humus have in no way been overrated.

It is a well-known practical fact that the more valuable classes of Oak and Pine timber cannot be profitably produced merely by repeated thinnings and by delay in the utilization of the crop [1], since the increased growing-space demanded by the crowns, and the consequent interruption of the leaf-canopy, are incompatible with the conservation of the productive capacity of the soil when it becomes exposed to the deteriorating influences of sun and wind. Strong thinning or partial clearance [2] of Oak crops about the sixtieth to the eightieth year, or of Scots Pine about the thirtieth to the fortieth year, after their chief growth in height has been attained on the different classes of soils and situations, with simultaneous underplanting or undersowing of the shade-bearing species best suited to the soil in question, is the only rational and profitable method of stimulating the trees to further production of good annual increment. Mere enlargement of the growing-space will of itself lead to enhancement of increment per individual stem; but, as has been above shown, its technical quality and its disposal will not be so good or remunerative, whilst this increment will ulti-

---

[1] Compare Loudon, *Arboretum et Fruticetum Britannicum*, 1838, vol. iii. p. 1809.

[2] When up to 15 % of the basal area of the stems is removed, it is a *thinning*, but beyond that it becomes a *partial clearance*; see Chapter IX on *The Tending of Woods*.

mately be dearly bought through the deterioration that must take place on any soils but those exceptionally circumstanced with regard to insolation and winds.

In the above the underwood has been dealt with only as regards its effects on the growth and development of the standard trees; it deserves, however, some consideration on its own account. It will practically be found that even when, at the time of the underwood being formed, all the standards appear sound and healthy, individual trees become sickly or lag behind in development to such an extent as to render their utilization advisable without attempting to retain them till the rest of the crop is mature. Where there is no underwood, this means that on such places the productive capacity of the soil is either not utilized at all, or at any rate not utilized to its full extent; but with underwood already formed, it can grow up into the blanks to be harvested along with the standard crop when this attains its financial maturity. In such cases Spruce, or Silver and Douglas Firs, although not in general attaining first-class dimensions, will yield better results than Beech or Hornbeam ; and, owing to the smaller growing-space requisite for the normal development of their crowns, they can often contribute a very fair return for the costs of formation and the sub-tenancy of the soil.

When the standards are mature, the underwood has fulfilled its duty, and may either be cleared away at once in order to admit of their natural regeneration, or may be retained longer and treated as high forest, at any rate in groups or patches showing good promise of fair outturn in timber. But, in accordance with the purely actuarial principles which should mainly guide the owners of woodlands throughout Britain, and in view of the comparatively small and unremunerative market for fire wood, the clearance of the underwood will usually precede the reproduction and utilization of the standards. Hence, for the underplanting of the deciduous, light-demanding species of our forest trees—Oak, Ash, Maple, and Sycamore—the Beech, and on moister localities the Hornbeam, are on the whole preferable to Spruces or other Firs, for they accommodate themselves more to the natural habits of growth of the standards. But, on the other hand, for Larch and Pine woods, though Spruces, or on the better classes of soil and in mild sunny situations Silver Firs, naturally conform more to the habits and development of coniferous standards, an underwood of Beech or Hornbeam is distinctly preferable when the conditions of soil and situation permit of it, owing to the great advantages conferred by them in respect of decreasing the dangers arising from insect enemies and fungoid parasites. In this matter the effect of the underplanting resembles the advantages gained by an admixture of broad-leaved species with the conifers ; and the benefits become more pronounced as the underwood grows up nearer the level of the crowns of the standards. Very often, however, the choice of the species to be used for underplanting depends mainly on the nature of the soil and situation, and is practically limited to the four great shade-bearing genera, Beech, Hornbeam, Spruce and Silver Fir — leaving out of consideration such exceptional cases as favour coppice-growth of Alder under essentially light-demanding species on moist, fertile, rather marshy soils. Among exotics of comparatively recent introduction Nordmann's Fir seems best endowed with the requisite shade-bearing and soil-improving capacity. If the Douglas Fir also prove suitable, the timber of the latter holds out very enticing financial promises to far-seeing landowners.

Of these genera, Spruce is the one which makes most moderate demands as to the quality of the soil; while it has also the advantage of confining them to the upper layer of soil through which the root-systems of deeper-rooted trees like Oak and Pine do not ramify. The other shade-bearing genera not only make greater demands for nutrients, but also tend to compete with the standard trees in extracting them from the deeper layers of soil. Speaking practically, Beech and Hornbeam deserve the preference if the protection and improvement of the soil is the main object in view; whilst Spruce and Silver Fir recommend themselves where it is desired to obtain fairly marketable timber from the underwood as well as from the standards. An underwood of Beech is productive of the greatest improvement in the soil, owing to the rich fall of its dense foliage and the fine quality of the humus formed therefrom. But, owing to the demands it makes on potash and phosphoric acid, it cannot be grown on the poorer classes of soil (just where it is most wanted); and in damp localities under Oak it is more apt than Hornbeam to suffer from late frosts in spring, owing to its flushing into leaf about a fortnight sooner than the standards. Where it is only advisable to underplant with Spruce or Silver Fir, or where these may be admixed with Beech or Hornbeam to form undergrowth, the Silver Fir generally deserves the preference, even although its roots strike deeper and its demands on mineral strength in the soil are greater; for it has on the whole a greater capacity for bearing shade, forms better humus, and is not so apt to close the soil to aeration and atmospheric precipitations, or to hinder the percolation of water from the surface downwards to the sub-soil. And, besides this, it has greater recuperative power later on when the canopy over head becomes interrupted from casual clearances of diseased or sickly stems among the standards.

For the underplanting of Oak, Spruce is not to be recommended, as it tends later on to interfere with the formation of the crown. Beech, or on moister situations Hornbeam, should form the major portion of the underwood, with Silver Fir scattered in patches here and there wherever there is any likelihood of its being allowed to develop into marketable dimensions. Most of the eminent sylviculturists in Germany condemn the use of Spruce as undergrowth to Oak, and urge, as chief of the several reasons, that, by intercepting and utilizing a considerable percentage of the atmospheric precipitations, it induces *stag-headedness* among the standards, owing to insufficiency of moisture in the layer through which the roots ramify, and from which they extract their food-supplies. This may, however, not infrequently be due to injudicious thinning or partial clearance on soils that are below the average in quality and are naturally rather deficient in moisture—though the evil is then, of course, aggravated by the selection of Spruce as underwood.

Although endowed with considerably greater capacity for bearing shade than Scots Pine, the Weymouth and Black Pines are not naturally adaptable for undergrowth ; far better results are promised in this direction by Nordmann's Fir and perhaps also by the Douglas Fir, which in many respects correspond closely with the Silver Fir.

Although the richest fall of dead foliage takes place when the whole of the standard crop is underplanted, yet the advantages of partial underplanting, with reference to decreased competition with the standards for the nutrient salts held in solution by the soil-moisture (for it is only when soluble that they can be imbibed by the rootlets), are so apparent as to make this method preferable on soils of only average or inferior quality. Because, whilst extracting smaller supplies of nutriment from the soil, it permits of a larger percentage of the aqueous precipitations reaching the soil, and of a somewhat speedier decomposition of the dead foliage than under close canopy, without endangering the quality of the soil through insolation and wind. That, by arranging the underwood in patches, better opportunities are given for studying the special character of each particular piece of ground, and for selecting the shade-bearing species best suited for it, with the additional advantage of attaining a mixed growth of underwood, need hardly be commented on ; nor the fact that, for the formation of undergrowth in patches, only a much smaller outlay will usually be requisite than if the whole area be planted up or sown.

Whether the underwood should be formed by sowing or

planting depends mainly on local circumstances. Sowing is usually cheaper, notwithstanding the fact that some measure of preparation is always necessary to prepare the soil for the reception of the seed ; but the object in view then takes longer to attain than when planting is undertaken. Wherever practicable, planting can be carried out with seedlings from neighbouring crops, or with two-to-three-year-old seedlings or transplants from temporary nurseries previously formed near at hand. It should not amount to more than about 15*s*. per acre, if the soil permits of notching, and if the plants are not put out closer than 4×4 ft. or $3\frac{1}{2} \times 3\frac{1}{2}$ ft.

Oak, Larch, and Pine are the kinds of trees extensively grown in pure forest in Britain, which stand most in need of underplanting when once their growth in height has culminated, and the individual stems have begun to assert their demands for larger growing-space. The other light-demanding species, like Ash, Elm, Maple, Sycamore, &c., are less frequently grown in pure crops than as subordinate patches in mixed woods where the ruling tree (which should usually be the Beech) can protect the soil against deterioration.

And even if the costs of the formation of undergrowth under pure woods of Oak, Larch, or Pine should locally exceed the above-named sum, the advantages of the method are cheaply obtained where the intention of the landowner is to obtain the best financial returns from forests that are unable to protect and enhance the productive capacity of the soil for themselves, but which he does not wish to utilize at present so as to make way for other timber crops. Many an Oak grove throughout England, and many a Larch and Pine tract in Scotland, would yield far more satisfactory returns from every point of view— financially, sylviculturally, aesthetically—if proper measures were taken for the protection and improvement of the soil by means of thinning, or partially clearing, with simultaneous formation of underwood and subsequent treatment of the crop in accordance with the rational principles of scientific Forestry.

Brief Recapitulation.　The Munich investigations showed that it has not yet been proved that undergrowth exerts any special and direct influence on the actual rate of increment of light-demanding standard trees.　But it unquestionably improves the shape of the bole, by tending to make the stem less tapering and therefore of greater technical and financial value, and also the quality of the timber (at any rate so far as concerns the light-demanding conifers) ; whilst by increasing the quantity of dead foliage and preventing its being blown about by the wind, it greatly favours the formation and increase of humus, and thereby materially improves the physical properties of the soil, and enhances its productive capacity.

CHAPTER XII

THE CONSERVATION OF THE PRODUCTIVE CAPACITY OF WOODLAND SOILS

No kind of soil can be said to possess inherent productive energy. The productivity or productive capacity of any land for agricultural or sylvicultural utilization is a mere potentiality ; whilst the light and warmth of the sun are the kinetic influences which set this in action, and enable it to sustain organic life.

Only those kinds of woodland crops can be grown prudently and *economically*, which merely utilize, but do not exhaust, the productive capacity of the soil,—which in fact, to use a financial simile, live on or within the annual income obtainable from the land, and do not encroach upon its productive capital. Whenever the form of crop is such as fails to safeguard the productive capacity of the soil, any immediate advantage promised by its retention is dearly bought at the cost of ultimate deterioration of the land to a greater or less extent. The rewooding of such deteriorated soils is often an extremely difficult task, more especially when they are of a decidedly limy nature.

I. Maintenance of, or Increase in, the Quantity of Mineral Nutrients available throughout the Soil in the form of Soluble Salts.

So far as concerns the actual withdrawal of sulphur, potash, magnesia, lime, iron, and phosphoric acid—the most essentially important constituents required for the growth of trees—it is seldom the case that any given soil is deficient in any of these minerals; although it sometimes happens that it has not sufficient moisture to enable it to offer them readily to the rootlets in the shape of soluble salts, in which form alone they can be imbibed by the suction-roots. With respect to the demands they make on fertility, woodland crops differ essentially from agricultural crops, and especially from the more exacting kinds like tobacco, hops, and sugar-beet, which all demand the application of manure in order to maintain the productive capacity of farm-land at par, and to prevent its rapid exhaustion and deterioration. In the majority of agricultural crops (hay, clover, lucerne, barley, wheat) from 4 to 8 % of the dry tissue remains behind as mineral ash on its combustion, and this even rises to 17 % in the case of tobacco, hops, and sugar-beet; whilst for high-forest crops of timber the total requirements *per acre* and *per annum* vary, according to Ebermayer's investigations, with the different species of trees from 4 to 20 lbs. as regards lime, from 2 to 10 lbs. as regards potash, and from about 1 to 4 lbs. as regards phosphoric acid[1]. Say, for example, that any given land contains only ·05 % of potash in a soluble form available for absorption throughout a soil of 18 to 20 inches in depth; then the total amount available per acre would be over a ton and a half, or, roughly speaking, about 3,300 to 3,500 lbs. per acre (taking the specific gravity of the earth as 1·5). But as, for Silver Fir, the tree which takes up most potash from the soil, only about 10 lbs. per acre are annually required for the formation of timber, even pure forests of this tree, if worked with a rotation of 100 to 120 years, would only have withdrawn from 1,000 to 1,200 lbs. by the time they attain maturity. They would therefore still leave the soil with an abundance of soluble potash for any kind of subsequent timber crop, even if in the

---

[1] Article in *Forstlich-naturwissenschaftliche Zeitschrift*, 1893, p. 227. See also Chapter IV on *The Nutrition and Food-supplies of Forest Trees*, p. 83.

meantime no fresh supplies had become available through the continuous decomposition of the soil-particles, or if these had sufficed merely to balance the annual loss due to soluble potash being washed away out of the soil by the percolation of moisture. And if this be the case with Silver Fir, it must *a fortiori* be true of all other species of woodland trees, and for all mixed forests, so far as potash alone is concerned. A similar argument likewise obtains with regard to the various other mineral food-supplies requisite for the growth of forest trees, either in pure crops or in the preferable form of mixed woods. Hence the conclusion must be arrived at that, wherever soils obviously deteriorate under timber-crops, the reasons should first of all be looked for in the physical conditions involved, rather than in any excessive demands made by the crops with regard to mineral nutrients.

So far as concerns any *direct increase* in the quantity of mineral food supplies contained in an available (soluble) form within the soil at any given time, the nature of the woodland crop, and the treatment to which it is subjected, can of course exert no influence. Great, however, is the influence which may be *indirectly* made to act upon the quantity of nutrient salts available for the rootlets, by maintaining such constant density of canopy overhead as will break the violence of heavy rains that might otherwise wash away the surface-soil, and as will also prevent the soil-moisture from being rapidly evaporated by sun and wind.

As most of the mineral food-supplies withdrawn from the soil are, during the process of assimilation, concentrated in the foliage, it follows that year by year the greater portion of what has been withdrawn by tree-crops is restored again to the soil when the decomposition of the leaves takes place. For when *humus* is being formed from the defoliated leaves and other dead parts of the tree-crops thrown down to the ground, carbonic acid and ammonia are given off. The former of these is essential to the solution of the mineral nutrients so as to render them available for the rootlets ; whilst the latter is itself an indispensable nitrogenous food. But this direct action of humus is in reality of secondary importance to its enormous influence in improving the physical conditions of soil, more especially with respect to the regulation of the soil-temperature, and with regard to its power of absorbing and retaining moisture, in preventing the rain from rushing off the surface and in allowing it to permeate gradually into the subsoil.

Though of itself affording no index to the mineral fertility of the land, a favourable admixture of humus throughout the upper layers of soils that are not of themselves fertile goes far towards making them suitable for the more exacting kinds of timber-crops. It is only when the normal progress of the decomposition of the leaves and fallen débris is interfered with by injudicious treatment of the crop—such as the maintenance of an excessive density of leaf-canopy in crowded plantations, or by allowing the leaves to be removed for litter, or to be blown away by the wind—that there ever is any practical danger of the humus and the upper layer of dead leaves having an injurious effect upon the productive capacity of the woodland soil which they cover.

But as a matter of fact, when the formation of humus goes on under the most favourable combinations of moderate warmth, moisture, and atmospheric oxygen—whose action is stimulated by fungi like *Cladosporium humifaciens*, Rostr.,—the decomposition takes place so continuously that but a slight covering of leaves shed during the past, and partially also during the previous, autumn litters the ground. Where the soil is exposed so freely that the benefits, which would otherwise be derivable from the humus, are diverted from the tree-crops and unprofitably consumed by a soil-covering of grass or weeds, care should, in the interests of the productive capacity of the soil, be taken to maintain a thicker canopy of foliage overhead, and thus keep down the rank growth of weeds by cutting off the supplies of light and warmth requisite for their growth. But where, on

the other hand, a deep layer of undecomposing leaves litters
the ground thickly, it prevents air from entering and circulating
within the soil, and induces an unfavourable condition with
regard to the moisture ; whilst it also tends to the development
of certain kinds of free acids of an injurious nature.   Such
conditions indicate either wetness or want of aeration in the
soil, or an excessive density of canopy overhead.   In our damp
insular climate, however, there is much less danger of the
productive capacity of the soil becoming endangered from this
cause, than from the opposite extreme, of having the timber-
crops more open and of a lighter canopy than is advisable for
the prudent husbanding of the means of stimulating the increment in the present crops, and for ensuring that those which
may succeed them will enter upon a heritage able to supply
the crops with sufficient supplies of nutriment.

Various tree-crops show different results with regard to the
capital in nutrient salts that they leave behind them in the soil
when they have reached maturity and fall to the axe.   As has
been pointed out in a previous chapter (see p. 86), conifers in
general, but more especially the Spruce, and in even a more
marked degree the Scots Pine, make relatively smaller demands
on mineral nutrients than broad-leaved deciduous species,
both with regard to their foliage and their timber.   Hence
tracts of woodland can be greatly improved by the rotation
of a coniferous crop, when once the soil has become exhausted
and deteriorated either by the free play of sun and wind, or by
growing light-demanding, thinly-foliaged kinds of trees (like Oak,
Ash, Maple, Larch, and Pine, when once they begin to get
broken in canopy) without an admixture of shade-bearing, soil-
improving kinds like Beech, Spruce, and Silver or Douglas
Firs.   Under the good canopy maintained by all the evergreen
conifers (for even the light-demanding Scots Pine has a good
canopy till about the twentieth to thirtieth year, when its
special demands for increased growing-space make themselves
apparent) the rich fall of easily decomposed needles forms

S

a beneficial layer of humus. By means of the carbonic acid and the ammonia evolved during the process of humification, a stimulus is given to the formation of larger supplies of nutrient salts within the soil. These are not required for the annual production of foliage or timber; but as the density of canopy overhead, and the mossy covering which often clothes the soil under conifers, both combine in preventing the nutrients being wasted by a rank growth of weeds, or being washed out of the soil by the unbroken descent of heavy rainfall, it practically follows that the sum total of soluble nutrients is, at the end of a coniferous crop that has been prudently managed, considerably larger than it was at the time the planting with Pines, Spruces, or Silver Firs was undertaken.

Many tracts throughout Germany, that half a century ago bore fine woods of Oak and others of the nobler species of broad-leaved trees, are to-day covered with Spruce and Scots or Austrian Pine in consequence of imprudence in the management of the former deciduous crops, and of the want of knowledge (since dearly purchased) that where a good leaf-canopy is not maintained, an undergrowth is requisite for the safeguarding of the productive capacity of the soil, or for its improvement if signs of incipient deterioration have already made themselves apparent. And in England the same may be said of many portions of the old Oak woods in the State properties, the New Forest and the Forest of Dean, parts of which have had to be transformed into coniferous woods in consequence of neglect of the conservation of the productivity of the soil. In the low demands that it makes on mineral nutrients, in the power of accommodation it exhibits with regard to soil-moisture, and in its capacity for improving poor soil or recruiting those that are temporarily deteriorated from exposure to inimical influences, there is no tree of the forest that can be more safely relied on to yield yeoman service than the common Pine. On poor sandy land it can form a crop where other trees fail to thrive; and even on inferior classes of soil it may be made to give fair returns in timber, whilst in reality also making other payment for its tenure of the land by the betterment of the soil and the improvement of its productive capacity for the immediate future. Where land-improvements of this nature can be made with mixed crops, these have such estimable advantages of their own (see p. 118), that an admixture of species should be effected wherever the circumstances of soil and situation permit of it; but where the soil is, for the present, of such inferior quality that nothing but Scots Pine seems likely to thrive, it will very often lead directly and satisfactorily to the desired result, if the crops be tended properly, and if no accidents occur from fire, snow, insects, or epidemic outbreaks of fungoid diseases. A slight admixture of Austrian and Corsican Pines should usually, however, be tried experimentally.

For the re-wooding or the recuperation of limy soils that have become exhausted through the action of sun and wind— one of the most difficult of sylvicultural tasks—there is no better method than first of all putting them under crops of Black or Austrian Pine (*Pinus Austriaca*). Even this first step towards the re-attainment of the high productive capacity characteristic of good limy soils is often exceedingly hard to achieve. Excellent results have at times been gained—even on hot southern hill-sides exposed to the full blaze of the sun during a continental summer—by sowing Lucerne and Black Pine seed together, and allowing the crop of the former to decompose and humify on the ground in order to assist the latter in maintaining itself for the first year or two, until the young plants become established in the soil. When once this is successfully effected, the rest of the work is comparatively easy. From the time that such crops have commenced to form canopy, the improvement taking place in the soil not only becomes marked, but can easily be maintained and enhanced by proper treatment of the Pine, and by a rational choice and treatment of the species to form the succeeding timber-crop.

The increase which may thus be made in the total quantity of mineral nutrients stored up in an available form within the soil cannot, however, be said to be directly due to any inherent quality in tree-growth, but must be credited as primarily due to indirect physical causes which give the sylviculturist an opportunity of rectifying errors made during the past.

Whenever any woodland soil deteriorates, it is almost certain that one or other, and usually two, three, or even all four of the physical factors[1]—*Cohesiveness or Tenacity, Relation to Moisture, Relation to Warmth, and Depth*—have been affected in some prejudicial manner. These physical factors determine, in a far greater degree than its mineral composition, the capacity of any given soil for yielding to timber crops abundant supplies of the nutrient salts requisite for their normal growth, development, and reproduction. All these physical factors, indeed, act and react so closely and essentially on each other, as in point of fact to determine the general quality of any particular soil for present sylvicultural utilization. Whatever timber-crops, and whatever methods of treatment of these, may be calculated to improve the condition of any one of these factors, will therefore—provided always that it be not altered to such an extent as to prejudice the utility of any of the other three factors—be a step in the direction of conserving the productive capacity of the soil, and of stimulating this when it is capable of being enhanced.

As was explained at the close of the chapter on ' The Nutrition and Food-Supplies of Forest Trees ' (see page 88), **Coppice-woods** in general, and more especially those in which the crops consist of Ash, Elm, Lime, Oak or Willow, genera of trees making the highest demands on the soil for nutrients, tend more towards the exhaustion of woodland soils than **Copse or Coppice under Standards** in which only the under-wood is cleared away periodically at comparatively short

---

[1] See Chapter V on *The Characteristic Influences of the different Classes of Soil, &c.*, p. 102.

intervals, twenty to twenty-five years; while this composite form of crop is in its turn less conservative of the productive capacity than **High-Forest** timber crops.

The frequent deterioration of soils under coppice does not arise on account of these kinds of trees making high demands *per se* for food-supplies, although for young poles, shoots from stools, and stoles of sucker-growth considerably higher demands are continuously being made than would be the case if these were to be allowed to develop into young trees having a larger proportion of ripe wood to the foliage and bark in which the mineral ashes are chiefly deposited ; but it is mainly ascribable to the fact of the soil being laid bare to the exhausting influence of sun and wind every twelve to sixteen years on the hags being cut over for the harvesting of the crops of poles, bark, or small material.

So far, therefore, as considerations regarding conservation of the productive capacity of the soil are concerned, high-forests in general, and those formed of thickly-foliaged species in particular, or copses of standard trees having a good protective soil-covering of underwood, are certainly the forms of woodland crops most thoroughly satisfying **the first fundamental principle of Sylviculture**, viz. *that the natural productive capacity of the soil must be carefully conserved, in order that it may satisfy continuously and uninterruptedly all rational demands made on the land with regard to the production of timber or of other forest crops.* Both the quantitative yield and the qualitative outturn in timber, or in other forest produce, are dependent on the manner in which this fundamental principle is kept in view. If, on merely average or inferior land, such kinds of crops be formed, or methods of treatment be adopted, as imperil the productivity of the soil, then any extra returns that are promised in the immediate future must be dearly purchased at the cost of the inevitable ultimate deterioration of the land for timber-production. It would, in fact, be merely discounting the future productivity of the land. The demands for timber throughout the civilized world are steadily increasing annually; while at the same time the total area under forests is being reduced. More discrimination in the formation and the treatment of crops should, therefore, be exercised generally than has hitherto been the case even in those countries that are furthest advanced in the art of forestry, and in all scientific knowledge pertaining to Sylviculture.

II. Minor Climatic Influences affected by the nature of the Treatment accorded to the Timber Crops.

Whilst influences regulating the productivity of large neighbouring tracts of agricultural land can be affected beneficially by the formation of extensive woodlands throughout districts in which the annual rainfall is less than 40 inches [1], and can also be prejudicially altered through great total clearances of forest growth, yet even on a very much smaller scale such local climatic effects as **early and late frosts** are to a certain extent within our control. Just as, in the case of coniferous crops, the annual falls should, for the purpose of protection against wind, succeed each other in the opposite direction to the prevailing or most dangerous winds, so also in undulating tracts and uplands, indented with coombs likely to collect the cold, heavy air at night, consideration should be given to the regeneration or the planting up of the endangered tracts first of all, before similar operations are carried out round about them. If their planting or natural reproduction be left till that of the surrounding land is completed, then, when the young crops on these have reached the thicket stage of growth, the circulation of the cold night air becomes impeded; hence there is danger not only of the plants put out being killed off, but also of these spots becoming **frost-holes** exceedingly difficult to plant up again with any of the less hardy species

[1] Endres, article *Forsten* in the *Handwörterbuch der Staatswissenschaften*, Jena, 1892, vol. iii. p. 607.

of trees that may otherwise be best suited for the given soil and situation.

On hillsides, therefore, the felling and regeneration should take place from above downwards. On being cooled by radiation of warmth from the surface of the soil, the air becomes heavier and seeks a lower level; and if its progress down towards the valleys be hemmed or made to stagnate by young thickets, the plants forming these are unduly exposed to danger from late frosts in spring and early frosts in autumn. On level stretches of land it suffices if the young crops are fringed on one side at least by a lofty crop of trees ; for then the cold air can circulate freely among the lofty stems in that direction in place of being penned in by older, densely foliaged thickets of somewhat higher growth. Hence, in natural regenerations, it is an error to await the return of another seed-year when patches here and there are deficient in seedling growth. It is much more advisable to sow artificially, or to plant, than to expose the ultimate seedling growth on such patches here and there to subsequent danger from late and early frosts on account of their being somewhat less in height than the seedling growth around them. Similarly, when groups of trees whose seedling growth is likely to suffer from frost,—such as Oak, Ash, Beech, Maples, Silver Fir, and Spruce,—occur of nearly mature age scattered among similar crops that are fully mature, it is better to regenerate them along with the latter, than to run the risk later on of developing frost-holes, where only the hardier species of trees—like Birch, Elm, Aspen, Alder. Larch, and Pines—will be able to outlive the dangers constantly recurring for several springs and autumns until the new crops of young plants get their head well above the frost-line, below the level of which the chill night air stagnates. And the quicker such spots are planted up, of course, the less danger is there of frost-holes being formed.

In order that patches of this description may be utilized to their full productivity, it is requisite that, if they are apt to be moist, they should either be drained or else planted up with one of the genera of trees like Elm, Birch, and Larch, that are at once hardy and capable of transpiring freely so as to thrive on a moist soil.

III. The Manner in which the Cohesiveness or Tenacity of the Soil may be affected by the Methods of Treatment of Woodland Crops.

The productivity of any soil is dependent to no inconsiderable extent on its cohesiveness. The whole relation towards moisture and temperature is very directly and essentially influenced by the fineness or coarseness of the individual particles forming the soil. Whether the land be naturally binding and tenacious, like heavy clays and argillaceous limes and loams, or naturally light and porous, like many sandy soils, the maintenance of a good leaf-canopy throughout all the woodland crops is of immense benefit in preventing the former class of soils from becoming apt to get water-logged or else dried up at different seasons, and the latter class from responding too rapidly to changes in the atmospheric temperature, and from running the risk of having portions of the soluble nutrient salts washed out of the soil after heavy rainfall. Heavy soils, that are insufficiently protected, are apt to become gradually and imperceptibly caked in the upper layers, so that the penetration, not only of the rootlets of seedlings, but also of the atmospheric air, is rendered more difficult; whilst sandy soils rapidly sink below the freezing-point at night during the cold seasons of the year, thereby exposing seedling crops to late and early frosts, and become too easily heated during summer days, thereby leading to excessive loss of moisture through evaporation.

The great importance attached to aeration, or circulation of air within the soil, is very practically recognized in agriculture by the ploughing operations that are carried out annually. In sylviculture similar operations are out of the question, except in osier-holts, in nurseries, and in the preparation of patches or strips of soil during natural regeneration or artificial reproduction by means of sowing or planting. Nor are they required when once the young seedlings have firmly established themselves within the soil ; for the latter is not only aerated and mechanically loosened by the humification of the dead foliage and the cleaving and fissuring action of the roots, but is likewise protected from the soddening and setting action of rain by means of the leaf-canopy overhead.

Actual experience shows that, except at the time of the germination of seed, and the initial development of young seedlings, so long as good canopy is maintained throughout the crop no danger is run of the soil becoming set or hardened on the surface by rainfall, as is so apt to be the case in fields that lie fallow for any length of time. The humification of the dead leaves and twigs, and of the roots of individual seedlings, poles, or trees removed during the periodical processes of weeding, clearing, or thinning, not only serves to improve the other factors of influence as regards productivity, but most directly acts on the cohesiveness of the soil-particles, by tending to loosen binding soils, and to render more cohesive those which are of a light sandy nature. This natural process is of course aided by the action of the various organisms that are always to be found more numerously in soils in which humification is going on, such as *Cladosporium humifaciens*, Rostr., and other fungi, insects, earth-worms, &c.

If the thorough aeration of the upper layer of woodland soil could be periodically undertaken, there can be no doubt that the productive effect of this measure would be speedily noticeable in the more vigorous development and greater increment of the crops. But, except on a small scale in osier-holts, this is, for financial reasons, out of the question ; and besides this, it is open to considerable doubt if the quality of the timber produced would be so good as that which may now be grown without any such artificial stimulation.

Soil that is exposed to the action of sun and rain is apt to become rapidly heated, whereby the process of humification is hastened on ; and, at the same time, the upper layer is apt to cake or harden, whilst its nutrient salts are washed out into the lower soil, in consequence of which the surface-soil deteriorates. The best safeguards against such unproductive conditions are undoubtedly the retention of a good normal canopy overhead, and the prevention of the removal of dead foliage from the ground either by wind or by any other means.

IV. The Manner in which the Quantity of Moisture in the Soil may be affected by the Methods of Treatment of Woodland Crops.

Unless care be taken to preserve the most advantageous quantity of moisture throughout the soil, the productivity of the latter becomes greatly prejudiced. For if, on the one hand, too much moisture be allowed to remain in the soil, this remains cold and inactive, the process of humification is retarded, the solutions of the nutrient salts are weak and thin, and the vital activity of the individual trees is below what it otherwise would be if the soil-temperature of the layers permeated by the root-system were higher ; and, on the other hand, when there is an insufficiency of moisture throughout the soil reached by the roots of the trees, the mineral food-supplies, though perhaps contained *potentially* in a far greater quantity than requisite for even the most exacting kinds of trees, are often not available in that soluble form in which alone they can be utilized by timber-crops. The former condition, implying excess of moisture, can often be remedied by drainage, or by the choice of kinds of trees for crops like Alder, Ash, Willow, Maples, Elm, and in a less degree pedunculate Oak, Birch, Aspen, Larch, Spruce, and Weymouth Pine, all of which are species that require to transpire a considerable amount of water through their foliage before they elaborate the requisite amount of mineral food necessary for carrying on active vegetation and building up their ligneous tissue. The latter condition, however, implying an insufficiency of water for the transpiratory requirements of many of the above species, and even of the other less exacting kinds like Beech, sessile Oak, Silver Fir, and Douglas Fir, and generally involving along with this but scanty supplies of mineral food, necessarily limits the choice of trees from among which it is for the time being possible to form woodland crops. As has previously been shown (see page 79), deciduous trees make far higher demands on soil-moisture than conifers ; and among the latter the most suitable species for forming crops on dry soils are the Pines, more especially the Scots and Black Pines on sandy tracts, and particularly the Austrian Pine on land of a limy character.

In order that the most advantageous degree of soil-moisture may be retained continuously throughout extensive woodland areas, it is necessary that the soil should be protected as well as possible from becoming exposed to the drying and evaporating effects of sun and wind. This object may be best attained by a judicious choice of the trees to form the crop, by avoiding the formation of coppice-woods on soils tending to dryness, by the maintenance of good leaf-canopy without overcrowding (which would act most injuriously on the soil-moisture), by regenerating the woods naturally—in place of making large clearances or clear fellings of all the trees on the annual fall, to be followed by sowing a planting artificially,—and by taking the ordinary sylvicultural precautions for preventing the dead foliage from being blown away and the normal process of humification being interfered with—for, owing to its extremely favourable hygroscopic qualities, *humus* is of invaluable influence in absorbing, in retaining, and in regulating the disposal of the atmospheric precipitations throughout the upper layers of the soil.

The nearer any soil approaches to dryness, the more necessary, and at the same time the more difficult, it is to husband the soil-moisture. It is a fact that very many species of trees, like Oaks, Elms, Maple, Sycamore, and indeed all essentially *light-demanding* trees, are more intolerant of shade on soils from which they have difficulty in obtaining their requisite supplies of water for transpiration, than on those which yield them an abundance of moisture and food. Hence, on dry soils, there is not only greater necessity for maintaining close canopy overhead in the interest of the productivity of the soil, but also at the same time a much stronger natural tendency on the part of the trees to have thinly-foliaged crowns, and to strive after an increase in growing-space. Under these latter circumstances the only proper sylvicultural treatment that can be adopted is to cut out and dispose of the kinds of trees that soon attain their maturity (such as Birch, Aspen, Willow, Ash, Maple, Sycamore), thin out individual trees of the nobler genera not likely to develop ultimately into valuable stems, and then underplant the Oak, Larch, or Pine retained with the species of underwood best suited to the given soil and situation. By this means the soil is not only afforded a good mechanical protection against the evaporating and drying effects of sun and wind, but considerable masses of dead foliage are thereby provided for humification, and for the improvement of the soil generally (see last Chapter, page 243). It may here be remarked that the Spruce is less adapted for forming the underwood on dry soils than any of the other shade-bearing species (Beech, Hornbeam ; Silver, Douglas, and Nordmann's Firs), for its shallow roots are apt to interlace and form a network throughout the upper layers of the soil, and to intercept the aqueous precipitations, thus hindering the percolation of a fair proportion of the water down into the lower soil, whence the deep-seated roots of the standard trees draw by far the greatest portion of their supplies of food and moisture.

Where woodlands are exposed to the action of dry winds, care should be taken to favour the growth of branches down to the very ground along all the outer lines of trees, so as to protect the dead foliage and prevent evaporation of the soil-moisture as much as possible. In coniferous crops a fringe of Spruce, Douglas Fir, Silver Fir or Black Pine, is to be recommended as forming the thickest mantle of foliage towards the open; but in crops of broad-leaved trees a very effective protective mantle can be formed by cutting the first few rows of trees back to the stool, and letting them spring up as coppice. Plantations in fields, or bordering on fields, should be fenced in with good thick live hedges. On dry hill-sides, and more particularly on those of a sandy nature and little retentive of moisture, care must also be taken not to make large clear annual falls of timber, but to regenerate naturally so far as possible. When once the seedlings have established themselves, however, the parent standards have to be rapidly removed; for on such inferior situations the young growth can ill bear any overshadowing that deprives them of the beneficial dews at night. On such localities a certain amount of soil-preparation for the reception of the seed is almost a necessity, as without that the seedlings find difficulty in establishing themselves, and are apt to be delicate. When this soil-preparation takes the form of horizontal rills along the hill-sides, it is more favourable towards water-catchment than any other method; consequently it tends to increase the amount of soil-moisture, and thereby to enhance the productivity of the land. Anything like **irrigation** of the soil can of course only take place in osier-holts, and in nurseries. But wherever it is feasible on dry hill-sides or on slopes covered with oak-bark coppices, a good deal of benefit may be derived at a comparatively small outlay by leading off water from any water-courses or ditches into small rills formed almost horizontally along the slopes by means of a light plough.

In order to prevent light sandy soil near the sea-coast from being blown about so as to form dunes of shifting sand, and in order to hinder the formation of a thin impermeable subsoil layer of moorpan by dry heather mould, quartz-sand, and a slight admixture of oxide of iron, the retention of a fairly good canopy and the preservation of all the foliage for humification are essential. When stretches of moorpan soil have been reclaimed by subsoil ploughing and planting up with Scots Pine, the undecomposed layer of heather-humus is often turned into mould of a more ordinary nature. And on reclaimed dunes, the layer of dead needles under the Pine trees not only acts as a mechanical hindrance to the lifting of the sand, but yields to it the beneficial humus that helps to bind it and to keep it moist. Such light sandy soils are unfortunately too poor to bear densely foliaged crops, or to permit of any underwood being formed with a view to the improvement of the soil.

V. The Manner in which the Relation of the Soil towards Warmth may be affected by the Nature of the Crop or the Method adopted for its Treatment.

The manner in which woodland soils respond to variations in the atmospheric temperature is determined more especially by the quantity of moisture contained in them, than by the specific warmth of the different kinds of soil. Colour, Conductivity, Porosity, Composition, and other physical factors all exert direct influence in this direction ; whilst the nature of the soil-covering is also of no little importance. That the quantity of moisture contained is of so much influence may be easily understood from the great latent power that has to be overcome in effecting changes in the temperature of water. And as regards colour, soils prove no exception to the law of radiation and absorption, viz. that dark bodies respond more quickly to variations than those of a similar texture but lighter in colour.

The more porous the soil, the larger the individual particles, and the greater the extent to which stones occur throughout the finer earth, the higher is the general conductivity of the soil.

Soils which are retentive of moisture, like clay and stiff loams, remain cold and inactive in spring. Porous, sandy, and gravelly soils become easily warmed, but are liable to the danger of radiating their warmth too rapidly, and are thus apt to sink below the freezing-point at the most critical times in spring, and again in the autumn to a less important extent.

Even a small admixture of moisture in sandy soils makes them less apt to respond closely to sudden diurnal and nocturnal changes in the atmospheric temperature ; hence any sylvicultural measures likely to increase the amount of moisture in light porous soils are beneficial. And in the same way whatever tends to remove anything approaching to excessive moisture in heavy land likewise stimulates the root-systems ramifying throughout it to a more active imbibition of nutrients ; for the rootlets are stimulated into activity far more by the soil-temperature than they possibly could be merely by the atmospheric warmth that is conducted down the stem. Apart from such extreme cases of wetness as make drainage advisable to a greater or less extent, it will be at once recognizable what an important influence may be exerted both on light and on heavy soils by an admixture of humus. Humus is strongly hygroscopic in absorbing and retaining moisture It acts as a non-conductor, not only in keeping out sporadic warmth from the lower layers of soil early in spring before the warm changes in the weather become settled, but also in retaining it well into the autumn when diurnal variations again become marked. It is, however, needless to recapitulate its good qualities, as they have already been commented on in each of the previous sections of the present Chapter. A good admixture of humus in the upper layers of soil, and a soil-covering of foliage undergoing the process of humification, will therefore, by tending

to keep the soil fresh or at most slightly moist, and by causing
it to respond gradually to actual periodic variations, and not
to mere sporadic daily changes of temperature, prevent the
commencement of the annual vegetative activity till the general
warmth is greater and the assimilative process can be carried
on with greater energy, and without risk of the young flush
of leaves and shoots being killed by late frosts. But the
advantages derivable from delaying the early awakening of
vegetative activity in timber crops have previously been considered [1]; and it is only necessary to point out here that the
relation of soil towards warmth will be most favourable when
the timber-crops are of such a nature as to favour the formation
of a good layer of humus, and when they are subjected to such
methods of treatment as will not interfere with the normal
process of humification. Thus soils naturally of a moist description can most advantageously be utilized for crops of Ash,
Alder, Oak, Elm, Maple, Sycamore, Willow, or Birch, where
the more densely foliaged species of trees, with their heavier
fall of dead foliage, might be apt to litter the ground with so
thick a layer of dead leaves as to interfere with the progress
of humification and to impede the entrance of air into the soil,
or where, by their greater density of canopy and their lower
capacity for transpiration through their foliage, they might tend
to keep the soil too moist and inactive. On land of this class
the more lightly foliaged trees have not only a better leaf canopy
than on dry soils, but the action of the sun may be even
beneficial where there is any tendency to excess of moisture.
On dry soils, however, there exists a necessity for the maintenance of good canopy, for admixture of Beech along with Oak,
and of Douglas Fir, Silver Fir, or Black Pines along with Larch
and Scots Pine, or for underplanting when the crop of timber
trees is unable for itself to maintain the soil against the action
of sun and wind, and unable to prevent the upper layers from

---

[1] See also chapter on *Underplanting*. pp. 236, 237.

being robbed of their soluble salts by weeds and useless growth.

VI. The Manner in which the Depth of the Soil may be affected by the Nature of the Crop, and the Method adopted in its Treatment.

So far as sylvicultural operations are concerned, but little can practically be done with a view of directly increasing the depth of the earthy soil throughout woodland tracts.   The natural processes of the disintegration and decomposition of rocks, stones, &c. are gradual and slow.   Much can certainly be done to improve the productive capacity, as has already been mentioned with reference to the planting up of poor limy soils ; and there is no doubt that the gradual humification of the dead foliage not only adds to the depth of the upper layer of soil, but also favours the processes of further decomposition, which again lead to increase in depth.   In the majority of cases, however, the existing depth of the soil is taken as one of the concrete factors, to which the choice of the crop and of the method of treatment must accommodate themselves.

Wherever the sylviculturist has to deal with shallow soils, he must naturally select species of trees for cultivation that are able to develop normally within the depth available for the ramification of the root-system.   To attempt, for instance, to grow high forest of Oak on a soil of only a couple of feet in depth, with a subsoil of an impenetrable nature for the tap-root and the main side-roots, could not reasonably be expected to prove a success, however fertile the land might be.   If the land were good enough, and there were reason to anticipate better returns from such than from any other crop, then Oak-coppice might be cultivated for the sake of tanning-bark. Wherever the soil under timber is shallow, any measures must be avoided that may probably lead to interference with the amount of moisture contained in it, or with the normal process of humification of the dead foliage shed from the crop, and consequently with the stimulation of the formation of fresh nutrients within the soil. Thus, gradual natural regeneration under parent standards, though perhaps merely in groups or patches here and there, will prove much more· prudent than extensive clear-fellings laying bare large areas at one time ; and this precaution is all the more necessary on steep and on rocky slopes, where, in addition to the powerful inimical influences of sun and wind, there also comes the mechanical danger of the surface-soil being washed away by heavy rainfall.

All of the above considerations, individually and collectively, lead to the simple result that the fullest utilization of the soil and the most effective safeguarding of its productivity are obtainable by one and the same means, viz. *by the continuous protection of the soil through the maintenance of a good leaf-canopy of normal moderate density, and through the retention of the soil-covering of dead foliage for the beneficial formation of forest mould by the usual process of humification.* Where the timber-crops stand too thin, as seems to me the general fault in British woods, the productive capacity of the soil is neither utilized to its fullest extent for the formation of the longest and straightest boles of timber, comparatively free from branches, nor is it husbanded as it should be in accordance with the first fundamental principle of Economy in general, and of Sylvicul·ture in particular. Where, on the other hand, the timber-crops are too thick or crowded, nutrients may be drawn in excessive quantity from the soil, and squandered in the individual struggle for existence, instead of being utilized to anything like the best advantage in the more rapid development of the predominant individual stems ; and at the same time the process of humification of the dead foliage is interfered with. These dangers are of course greatest on poor lands that are either too moist or too dry. Just, in fact, where an admixture of humus is most requisite, its formation is most difficult.

A good canopy of foliage protects the soil from any excessive evaporation through the action of sun or wind, and helps to maintain the advantageous, non-conducting soil-covering of dead foliage, which gradually becomes of greater benefit to the soil itself and to the growing timber-crop as the process of humification advances. Hence woodland soils, that are judiciously utilized, provide for themselves the very best means of maintaining their productivity, so long as they are treated in such manner as will endow or retain them with the most advantageous degree of freshness or moisture; for along with this will, from the very nature of the factors upon which this is dependent, certainly be favourably combined the other physical properties and organic conditions which work together in order to produce the highest quantitative and qualitative returns obtainable from any given soil and situation. It stands to reason that these highest results can hardly be expected from the small patches, and clumps, and groves of woodland so often to be found dotted about large estates, more often really for the protection of game than for timber-production, but that they will be most easily and most effectively attained, the larger and the more compact the areas are, which may be set apart for the express purpose of growing timber in a rational manner in accordance with the Principles of Sylviculture.

CHAPTER XIII

THE FUNGOID DISEASES OF FOREST TREES, AND THEIR PREVENTION

No small proportion of the disturbances which take place in the growth of plants, and which vary from comparatively trifling ailments up to serious injuries resulting in the death of trees, is directly ascribable to the influence of parasitic, cryptogamous, vegetable organisms called **Fungi,** which live on or in the plants in question.

Fungi are plants of lowly organization, consisting only of cells and containing no chlorophyll. They consequently cannot elaborate organic substance for themselves, and therefore only obtain their requirements in this respect by withdrawing organic nutriment from other living or dead organisms belonging either to the animal or the vegetable kingdom.

Until the last few decades this pathological branch of phytology received scant attention, so far as woodland growth was concerned. Many of the fungal diseases of trees were unknown and contemptuously disregarded ; whilst, with respect to some others of a more prominent and widespread nature, cause and effect were confused. The appearance of many fungi, e. g. species of *Polyporus* causing red and white rot, was regarded as the consequence of the rot in place of its cause ; and not a few of the phenomena are still a sealed book to sylviculturists. The Pasteur of this branch of Forest Science is Professor Robert Hartig of Munich.

Fungi may be either **saprophytic** or **parasitic**. The former, fortunately including most fungi, are only to be found on dead and decomposing substances, and are therefore the accompaniment or consequence of sickly growth and disease, and not its cause. Those belonging, however, to the latter class, attack healthy living plants, and are a direct cause of disease, often terminating in death; but many of them continue in a sort of saprophytic growth when once the disease they have occasioned has terminated the life of the plant attacked. All the fungi called *rusts* (*Uredineae*) are parasitic.

The reproductive organs of fungi are developed in or on receptacles, sporangia, sporophores, or spore-producers, and consist of germinative cells of different kinds roughly classifiable as *Spores*, *Sporidia*, and *Gonidia* or *Conidia*.

Spores (including ordinary spores, and also Uredospores, Aecidiospores, Teleutospores &c., according to the different circumstances relative to their formation and development), are the generic name for these germinative cells without distinction; whilst **Sporidia** are secondary spores formed on a Promycelium developed by the germination of hibernating or resting spores; and **Gonidia** or **Conidia** are those formed at the point of usually erect-growing mycelial threads.

When any spores that settle on plants find the conditions (damp and warmth) favourable to their development, germination follows, i. e. they develop delicate, thin-walled, mostly colourless, tube-like processes (*mycelial filaments* or *hyphae*), sometimes filled with a golden-yellow oily fluid. These tubes, though often undivided, are as a rule divided into simple cells (septated), their contents being protoplasm and cell-sap. With the exception of the mildew-fungi, which merely vegetate on the exterior, the infection of leaves, fruits, bark and wood always commences from the outside by the hyphae pushing their way either through the stomata or the epidermal cells of young leaves or bark, or effecting an entrance into the interior of the plant through the unprotected portions of the roots, or at any wound-surfaces like those of broken branches.

These hyphae either insinuate themselves between the cellular walls of the parenchym or the prosenchym, and vegetate in the intercellular spaces or the resin-ducts, whilst they send single short off-shoots or suction-roots (*haustoria*) into the interior of the cells, or else they bore through the cell-walls, and thus force their way from one cell to another. By extending in length, and at the same time throwing out side-processes—less frequently by ramifying—they at length form a filamentous network or **Mycelium**, which represents the complete vegetative body of the organism. Masses of mycelium form **Sclerotia**, containing stores of food-supplies (chiefly proteids and oil) which, under favourable conditions, produce either new mycelium or sporophores of the fungus.

Receptacles or *spore-producers* spring from the mycelium and produce the reproductive organs, the spores. One and the same fungus often produces sporophores of the most varying shapes, and the form of these is much more characteristic (e. g. mushrooms, toad-stools, &c.) than the hidden mycelium, which is much alike in all fungi.

When the sporophores are only single threads from the mycelium they are called *spore-hyphae*, but in other cases the sporophores or receptacles are of the most varying forms, and have different names. *Carpospores* are spores which produce a second generation differing from the original form, whilst *Gonidia* or *Conidia* only reproduce the form on which they occur. Gonidia therefore multiply one form of fungus very rapidly within the annual period of vegetation, whereas carpospores perpetuate the species from one year to the other.

The spread of infectious diseases can take place either by mycelial infection from root to root under the ground (as in *Trametes pini*, *Agaricus melleus*, and *Rosellinia quercina* in plantations, and *Dematophora necatrix* in vineyards), or by spores and gonidia wafted about by the wind or conveyed mechanically by animals or other agencies (as in the case of *Phytophthora omnivora* on seedlings, new outbreaks of *Trametes radiciperda* at the base of the stem, and the rust on corn occasioned by species of *Puccinia*)[1].

The consequences of such infection with fungal hyphae are interference with the normal transpiration and the decomposition of the carbonic acid in the part infected, gradual destruction of the cells and consumption of the cellular substance and contents, with chemical disturbance and morphological alteration, generally accompanied by more or less hypertrophy or morbid enlargement, and resulting finally in death.   In consequence of the sickly state of the plant, insects are often attracted and assist in the work of destruction.   From the mycelium at length proceed the various forms of spore-bearing organs characteristic of the different kinds of fungi, breaking out and appearing sometimes on leaves, twigs, or on the surface of the bark, at old branch-holes, or even from holes through which bark-beetles have previously made their exit from the stem. Millions of reproductive spores are produced, and the cyclus of generation begins anew.   Some fungal species of sylvicultural importance deviate from this normal type in having frugiferous organs (*Rhizomorpha*) without any fructification of the mycelium (e.g. in *Agaricus melleus*, a species closely allied to the common mushroom).

Many fungi complete their life-cycle within a few weeks or months, whilst in other species the spores hibernate, and in others again the mycelium retains its vital energy for two, three, or more years; most of the species occasioning diseases of forest trees belong to this last-named category.

The polymorphic tendency of many fungi, or the change of

---

[1] Smut on wheat is occasioned by *Puccinia graminis*, which in its aecidial form, occurs as *Aecidium berberidis* on the leaves of the Barberry. This shrub should therefore be treated as a very noxious weed in all hedges, &c., throughout corn-growing districts in which rusts and smuts are in the slightest degree prevalent.

generation, demands special mention. It not infrequently happens that the spores of one fungus do not produce an individual resembling the parent, but something which appears quite a different kind of fungus; and the spores of this again may perhaps produce what seems a third kind of fungus not resembling either of the two previous forms; but ultimately the spores of the second or third forms produce, or throw back to, the original parent type, and the cycle re-commences. Accurate observations have thus shown, and are still showing, that many fungi previously reckoned as belonging to different genera are merely different stages in the development of other well-known species. Those genera are of course most highly organized, in which sexual processes can be proved. The regular change of form is called *change of generation*; the change in the shape of the regenerative organ is known as *pleomorphy*; and the change in the choice of a *host*, i. e. of the plant infected, is termed *heteroecy*. This change of generation finds its complete parallel among certain gall-wasps (*Cynipidae*) in the insect world. Thus *Cynips aptera*, producing galls on Oak-roots, gives birth to *Cynips terminalis*, forming galls at the ends of the twigs, whose progeny throws back to the original form *C. aptera*, and the cycle commences anew.

The occurrence and the spread of fungal diseases are specially favoured by certain conditions of climate and situation. Dampness and warmth are requisite essentials, whilst light is of less importance, as their peculiar processes of assimilation proceed without the agency of chlorophyll, inasmuch as they cannot, like green plants, assimilate carbonic acid; hence, if light were essential, they could not grow underground, and in the inside of timber trees, as they do. In damp seasons, and in humid localities with heavy, stagnant air, fungoid growth is more favoured than in dry years, or in exposed situations with free play of winds. Nitrogenous substance favours their growth, as the plasmic contents of the mycelial threads contain a large proportion of Albumen. In timber-work, the quantity of nitrogenous matter contained in the wood determines its liability to decay from fungoid causes (*Merulius lachrymans*, &c.) ; hence Beech or Willow wood is far less durable than Oak or Pine. But if there were no albuminous substance in woody tissue, or if it could be kept absolutely dry, fungoid decay would not take place in wood-work ; for moisture is an absolute necessity to fungoid growth.

In woodlands, fungoid disease may sometimes be noticeable as epidemic or somewhat general simultaneously throughout the crop, otherwise it may be merely sporadic or distinctly localized in places here and there. In the former case it is due to conditions closely connected with soil and situation or with climatic factors unsuited for the species of trees forming the crop, and can usually only be prevented by change of crop to a more suitable species ; whilst in the latter case it can generally be eradicated by excision, root and branch, of the infected individuals so as to obviate the spread of the disease.

Within the short space here at disposal it is impossible to undertake the detailed descriptions of any but the more important and frequent fungal diseases of our forest trees ; and, in considering them, it will be advantageous to discard any attempt at the natural classification according to reproductive peculiarities in orders and families, and merely to group them together according to the general tendencies they show with respect to the infection of *Foliage*, of *Stem and Branches*, and of the *Roots and the base of the Trunk*.

It may be remarked, generally, that (as will also be seen to be the case as regards injurious forest insects) fungoid diseases do much more damage in coniferous crops than in woods formed of broad-leaved species. This may be accounted for not only by the natural attraction they seem to offer to the different kinds of fungi, but also partly on account of the much weaker recuperative power the conifers possess in comparison with deciduous, broad-leaved species of trees.

General Classifications of the Chief Fungoid Diseases of Sylvicultural Importance.

I. On Foliage.

A. Of Conifers :—

1. *Hysterium pinastri*, Pine scab or scurf. Mainly attacking Scots Pine.

2. *Chrysomyxa abietis*, Spruce needle-rust or blight. Mainly attacking Spruce.

3. *Aecidium pini*, var. *acicola*, Pine needle-rust or blight. Mainly attacking Pines.

4. *Trichosphaeria parasitica*, Silver Fir needle-blight. Mainly attacking Silver Fir and Spruce.

Minor disorders are occasioned by *Hysterium nervisequium* on Silver Fir, *H. macrosporum* on Spruce, *Caeoma laricis* on Larch, *C. abietis pectinatae* on Silver Fir, *Aecidium columnare* on Silver Fir, *A. abietinum* on Spruce, and *A. strobilinum* on the bracts of Spruce cones.

B. Of Broad-leaved Trees :—

1. *Phytophthora omnivora*, Beech-seedling fungus. Mainly attacking Beech, Maples, Ash ; Spruce and Scots Pine (in nurseries and young seedling growth).

Minor disorders are due to *Melampsora Hartigii* on Willows (especially *S. caspica*) ; *M. tremulae* on Aspen, *M. Betulina* on Birch, and *M. salicina* on Sallow, also *Rhytisma acerinum* on Maples and Sycamores [1].

---

[1] It may here be remarked that there is reason to believe that *Caeoma laricis* and *C. pinitorquum* can both of them pass through their aecidial stage either as *Melampsora tremulae* on Aspen or as *M. Betulina* on Birch ; and that the spores of either of these latter forms may infect the shoots of Scots Pine with *Caeoma pinitorquum* or the foliage of Larch with *C. laricis*. Hence Birch and Aspen should be treated as weeds and cut out, wherever either of these diseases is prevalent among Pine or Larch woods.

II. On Stems and Branches (in the bark or in the wood).

The cankerous diseases which belong to this section may be divided into two main groups. In the first of these are comprised all such diseases of the bark and the cambium as cannot be healed without human interference, because of their influence annually extending, so that when it has encircled the stem all above this zone dries and dies off. And in the second group are comprised disorders which only, as a rule, spread through the bark during a single period of vegetation; but if within that time the disease has not completed its circuit round the stem the infected place becomes cicatrized and heals over. Unfortunately the only one of the chief cankerous diseases below mentioned which belongs to this latter group is *Nectria cucurbitula*.

A. Of Conifers :—

1. *Trametes pini*, Pine fungus. Mainly attacking Scots Pine, Spruce, Larch, Silver Fir.

2. *Aecidium pini*, Pers. var. *corticola*, Pine-canker or bark-fungus. Mainly attacking Pines.

3. *Caeoma pinitorquum*, Pine-shoot fungus. Mainly attacking Pines [1].

4. *Aecidium elatinum*, Silver Fir fungus. Mainly attacking Silver Fir.

5. *Nectria cucurbitula*, Spruce-bark fungus. Mainly attacking Spruce.

6. *Peziza Willkommii*, Larch canker. Mainly attacking Larch.

Minor disorders are occasioned by *Cladosporium entoxylinum* and *C. penicillioides* on Scots Pine, and *Pestalozzia Hartigii* on Spruce and Silver Fir (seedlings in nurseries).

[1] See foot-note on opposite page.

B. Of Broad-leaved Trees :—

1. *Nectria ditissima*, Beech canker. Mainly attacking Beech, Oak, and all other similar trees (except Birch?).

Minor disorders are due to *Nectria cinnabarina* on Maples, Elm, Lime, and in particular Horse Chestnut, and *Polyporus sulphureus* on Oak, Poplar, Tree Willows, Birch, and Larch.

III. On Roots and Base of Trunk.

A. Of Conifers :—

1. *Agaricus melleus*, the Honey fungus, an edible mushroom. It mainly attacks conifers, but especially Spruce, or Scots and Weymouth Pines.

2. *Trametes radiciperda*, Root-fungus. It mainly attacks Scots and Weymouth Pines, Spruce, and Silver Fir.

B. Of Broad-leaved Trees :—

1. *Rosellinia quercina*, Oak-seedling fungus, occurring in one to three-year-old Oaks principally.

Details with regard to the Appearance, Causes and Effects of the Chief Fungoid Diseases, and the Preventive Measures adoptable against their spread.

I. Fungoid Diseases of the Foliage.

A. On Conifers :—

1. *Hysterium pinastri*, Schrad., the **Pine scab or scurf**, is one of the principal causes of *leaf-shedding* in young plantations. During late summer and autumn small brown spots, which contain the characteristic mycelium of this fungus, make their appearance on the young foliage of Pines. Early in the following spring the needles wither, redden, and drop off. If the winter has been mild and the spring is wet, the blackish fruits or

apothecia are developed in May ; but they can only burst when favoured by long-continuous rain. Hard winters and dry spring weather therefore tend to prevent the spread of the disease. This disorder is of very frequent occurrence in nurseries and young plantations.

In Germany thirty years ago this disease of the Scots Pine was merely '*an interesting observation*,' but now it has practically become a widespread calamity[1]. Similar appearances of leaf-shedding in the Pine may, however, easily be occasioned by other causes. Although R. Hartig and Prantl have shown it to be due to *H. pinastri* in a great many cases, Ebermayer has proved that it can be caused by a process of drying up or exhaustion owing to excessive transpiration in early spring ; whilst Alers and Nördlinger proved that it was frequently due to autumn frosts in rapid contrast with sunny days owing to strong radiation of warmth from unprotected soils, more especially after wet, cold summers unfavourable to the hardening of the shoots.

This disease attacks Scots Pine chiefly, and also the Austrian Pine, between the ages of one to five years, although it is likewise found on plants up to about twenty years of age. Damp, misty localities favour its spread, hence it is more frequently found in plains and valleys, than in hill-side woods. The best measures adoptable to prevent its spread consist in the admixture of Spruce with the Pine, in the avoidance of delay in pricking out seedlings in the nursery beds, in the formation of Pine nurseries in places free from infection—amid broad-leaved woods if possible,—and in the burning of all plants infected.

Some similar disorders are occasioned by *H. macrosporum* on the two-year-old foliage of Spruce plantations from ten to thirty years of age, and by *H. nervisequium* on two-year-old and older foliage of Silver Fir.

---

[1] See Chapter VI. p. 130.

2. *Chrysomyxa abietis*, Ung., **the Spruce needle-rust or blight.**

Pale yellow spots break out here and there on the needles of one-year-old shoots about the end of May or the beginning of June, which increase and become of an intense yellow whilst the uninfected parts remain normally green. Towards autumn brown longtitudinal marks make their appearance along the underside of the needles on both sides of the midrib, which become ruddy and slightly swollen during the autumn. These beds of *Teleutospores* become enlarged in the following spring; the epidermis bursts along the middle of the swelling in April or May; and the spore-bed makes its appearance as a velvety, bright orange pustule, which scatters its spores about the middle of May over the new flush of needles. The green parts of the foliage attacked soon withers, and the needles fall off in June and July.

This fungus only attacks the young needles, hence one-year-old or older needles are exempt; and the upper portions of the crown are much less liable to the disorder than the central and lower portions. Spruce plantations of from ten to twenty years of age, in damp, close situations with southern and south-western exposures suffer most, although the disease is also found in woods of thirty to forty years old. Its spread is said to be much more prevalent on limy than on other soils. Warm, damp spring weather favours its increase, whilst all atmospheric influences checking the early flush of the foliage militate against it. The direct injury done to the plantations is only considerable when the disease has obtained a foothold on the foliage for several successive years; but indirectly it is of real importance, as it vastly increases the danger of attacks from Bark-beetles. The only means of preventing attacks is to avoid planting Spruce in pure forests on wet soils and close, confined situations; whilst remedial measures are confined to the removal of all twigs infected, and to the regular conduct of thinning operations.

Somewhat similar appearances of rust or blight are caused by *Caeoma laricis* on the needles of Larch, and by *C. abietis pectinatae*, and by *Aecidium columnare* on Silver Fir.

3. *Aecidium pini*, Pers., var. *acicola*, **the Pine needle-rust or blight.**

During April and May, on the one and two-year-old foliage of young Pine, minute orange-yellow blisters (*Peridia*) are often to be seen in a row, either on one or both sides of the needles, between which the *Spermogonia* lie. These turn brown when mature, as also do the *Peridia* before they burst in the middle and scatter their spores. The mycelium develops in the inside of the needle without killing it, and again produces *Aecidia* in the following year. The foliage is only killed off and shed when the disease infects the plants extensively ; in this case its marks may be traced by small, blackish, warty spots with light edging.

Although mainly confined to Pines (and in particular to Scots and Black Pines) of from three to ten years of age, it may be found in woods up to thirty years old, but always on needles of the last or of former years, and never on the new flush of foliage. Here again the formation of mixed in place of pure woods, careful clearing and thinning, the removal of infected twigs, and the removal of groundsel plants (species of *Senecio*) on which the other generative form *Coleosporium senecionis* is produced, constitute the only practical measures for obviating the spread of the disease.

Somewhat similar disorders are caused on one-year-old Spruce foliage by *A. abietinum*. To one of the same genus, *A. strobilinum*, is also due the fungoid disease common on the bracts of Spruce cones.

4. *Trichosphaeria parasitica*, R. Hrtg., **the Silver Fir needle-blight.**

The thin, colourless mycelium is generally to be found on the lower side of the last sprays of Silver Fir, whence it usually attacks the needles pointing downwards, forming close white pustules on the whitish streaks along the under surface. In consequence of infection the needles change colour, turn brown, and drop off from the twig, to which they remain suspended by the mycelium. The pustules turn brown after the drying-up of the needles; and about November small, round, blackish-brown, hairy *Perithecia* are formed containing the dirty grey spores, which germinate easily on Silver Fir foliage and thus spread infection. But the mycelium hibernates on the sprays and needles, and extends to the new shoots developing in spring, whose foliage it attacks as well as the needles that may hitherto have escaped on the old sprays. In addition to scattering abroad myriads of spores, this fungus never relaxes its attacks or quits the foothold it may once have attained on any young tree, until the latter has been killed. This disease occurs most frequently in twenty to forty-year-old crops of Silver Fir, especially on the lower foliage and on individuals of most forward development; but it is also to be found on Spruce. The best remedial and protective measures consist in the cutting out of infected twigs, and in careful tending so as to maintain a good canopy and keep the woods as free as possible from tangled growth.

B. On Broad-leaved Trees :—

1. *Phytophthora omnivora*, de Bary, **the Beech-seedling Fungus.**

The effects of this disease were first noticed in the cotyledonous leaves of Beech-seedlings. They manifest themselves in the withering of either the slender stalk or the rootlet, and in black spots at the base of the cotyledons or on the primordial leaves when these appear. Within about a week of the first symptons of disease the whole seedling exhibits its effects, especially when May and June are damp months. In dry seasons the seedlings infected have a blackened, scorched appearance.

Infection primarily takes place by means of *ovispores* (*oospores*) dormant in the soil. As the mycelium is formed, it spreads throughout the whole stalk as well as within the cotyledons; whilst numerous *hyphae* break through the epidermis from within, or force their way through the stomata, and develop into *sporangia*-bearers. After the detachment of the citron-shaped sporangia their receptacles become prolonged and again form sporangia; whilst the older ones drop off and either at once germinate, or scatter in all directions the Conidia which they contain. The development of this fungus is so rapid that in rainy weather, and in close, damp situations, new sporangia-bearers are formed within three to four days of the first appearance of infection. Ovispores (*oospores*) are at the same time formed by the sexual process of fructification. These become incorporated within the soil when the infected parts rot away, and retain germinative power for three to four years, or even longer. When once this fungus establishes itself in nurseries, or in areas being naturally reproduced, it may occasion very serious damage, as in addition to infection from ovispores (*oospores*) that are easily carried from place to place by wind and passing animals, the mycelial filaments produce infection subterraneously by entering into the rootlets and suction-roots of neighbouring seedlings.

Besides the Beech, the seedlings of Ash, Maple, and Sycamore are most liable to its attacks, as well as those of all conifers, but especially Spruce and Scots Pine. So long as the infection only takes place at the tip of the seedling, recovery is possible; but where, coming from below, it begins to make its appearance in the stem, the case is hopeless. In consequence of this fungus whole seed-beds covered with rills of coniferous sowings may be killed off even before the germinating seedlings have made their appearance above the soil.

Preventive measures include the careful removal of infected seedlings whilst the disease is as yet merely sporadic. Any seed-beds which have been attacked should only be used for the schooling of transplants during the two or three following years ; if possible a different species of plant should be pricked out from what has previously been sown there. Watering the seed-beds with a solution of 4½ lb. of copper vitriol (blue-stone) and one quart of ammonia in fifty gallons of water is said to yield good results in checking the ·spread of the disease.

Minor disorders are occasioned by *Rhytisma acerinum*—which causes the yellow spots (*Xyloma acerinum*, the summer form), that change in autumn and winter to black, on Maple and Sycamore foliage,—*Melampsora tremulae* on Aspen, and *M. Hartigii* on the leaves on the one-year-old shoots on Willows: this last disorder is often serious on *S. acutifolia* and *S. caspica*. A similar fungus, *M. salicina*, is also found on Willows, on *S. caprea* chiefly, and *M. Betulina* on the Birch. In all these cases the burning of infected foliage should be seen to during autumn and winter ; whilst in osier-hags infected shoots should be cut away and burned wherever the disease manifests itself.

II. Fungoid Diseases of the Stem or Branches
(in the bark or in the wood).

A. On Conifers :—

1. *Trametes pini*, Fr., the **Pine Fungus.**

This disease is to be found chiefly in Pine stems from about forty years of age onwards, although it also occurs on Larch, Spruce, and Silver Fir. The spore-tubes insinuate themselves into the woody tissues on wound-surfaces or wherever green branches have been broken off. When the germination of the spores begins, the *hyphae* push forward, destroying the cell-walls and forcing their way into the heart-wood of the tree. Here the mycelium develops, and forms ring-shakes or heart-shakes extending down from the crown and leading to the rapid decomposition of the timber in consequence of the rapid increase of the mycelial filaments, whilst as a rule the sapwood remains untouched. The mycelium next makes its appearance at branch-holes of the trees attacked, or else direct out of the bark, where it forms a brown, woody, bracket-like receptacle or *hymenium*; and the spores scattered from this infect other wound-surfaces, unless these are protected by resinous outflow, or by having been tarred after any pruning of the Scots Pine. The technical value of Pine stems often becomes much reduced owing to this disease. It often happens that these spore-producers live for a long time (up to fifty years, according to R. Hartig[1]) and gradually assume large dimensions.

The immediate removal of trees attacked by the fungus appears advisable, not only in order to be able to utilize the timber before decomposition has proceeded too far, but also in order to prevent the spread of the disease by the scattering of the spores produced in immense numbers in the receptacle. The removal of Scots Pine branches in which heart-wood has already begun to form should be avoided ; or, if pruned, then the wound-surfaces should be coated over with tar.

2. *Aecidium pini*, Pers., var. *corticola*, **the Pine Canker or Bark-fungus.**

This is a disease to which Pines in general, and Scots Pine in particular, are liable both as poles and trees, but principally as young poles about fifteen to twenty years of age. The fungus appears first on portions of the tree that are of at least two years' growth, mostly at the whorls, and near the top of the tree.

Some kind of wound in the bark is necessary before the spores of this fungus can effect an entrance ; but when once this has been gained, the presence of the disease is first noticeable by the appearance of small bright semi-spherical or oval orange-reddish pustules (*aecidia*) protruding from the bark of branches and stems in June, which swell up and burst, scattering the spores they contain. The colourless, septated *hyphae* of the fungus live intercellularly between the parenchym cells of the bark, the liber, and the medullary rays, and send *haustoria* or

[1] *Op. cit.*, p. 164.

U 2

short side-branches through the cell-walls into the inside, in search of nourishment.

When the mycelial filaments come in contact with the starch contained in the cells, this becomes transformed into turpentine, which collects on the inner wall and saturates the tissue itself, so that a process of resinification or inspissation of turpentine sets in and interrupts the circulation of sap throughout the portions affected. These not only extend far up and down the stem, but also reach to a depth of three to four inches in large stems. As the mycelium annually increases in size, the cankerous parts clogged with resin also increase; whilst the stems assume an eccentric form of growth owing to increment taking place only where the bark is not yet infected. Sometimes the upper portion of the crown is killed off within a year, when infection has taken place nearly all the way round the stem ; but at other times decades sometimes pass before the crown becomes dry and death ensues. Warm, dry summers favour the advances of the disease owing to the stimulus given to transpiration through the foliage, with which the partial supplies of sap yielded by the uninfected portion of the stem cannot maintain a due balance.

The spores scattered by the *aecidia* have a change of generation with *Coleosporium senecionis*, having species of *Senecio* as their host. The mycelium hibernates on these, and in the following spring develops new spores by means of which the canker of the Pine is produced.

When once this disease has been noticed, the best measures adoptable are to cut out all the stems attacked before the fungus has had time to develop fully, and to destroy, so far as possible, all two-year old species of *Senecio* before they flower in April; this must be done by rooting them out, otherwise the stalks removed are replaced by new stool-shoots.

3. *Caeoma pinitorquum*, de Bary, the **Pine-shoot Fungus**. This fungus is mostly to be found on Scots and Weymouth Pines of from one to ten years of age, although young crops

are not safe against attack until about thirty years old. It occurs principally on wet soils, and after cold damp springs.

The mycelium of this fungus develops itself in the green parenchym of the bark of the youngest shoots, within which it is perennial, and extends itself into the new shoots when they flush in May. Shoots thus taken possession of by the mycelial filaments show a brownish discolouration of the woody tissue right into the pith. Before the shoots have attained their full length, pale yellow spots, bearing within the bark-parenchym the spermogonia of the fungus, become apparent about the end of May or the beginning of June, from the middle towards the end of the shoots. With the development of the sporidia under the bark the light yellow spots turn reddish-yellow, and become raised like pustules until the epidermis at last fissures longitudinally, and the spores are scattered. Each such fissure produces slight hypertrophy; and, as the cellular tissue dies down as far as the woody substance, this necessitates a bend in the shoot, which becomes concave at the diseased part, and at the same time naturally shows an outflow of resin. If the damage is only slight the wound cicatrizes and in time becomes obliterated; but when eruptions of spermogonia take place, as often happens, on alternate sides, a peculiar twisted form of growth is characteristic of this disease. When damp weather in May and June favours the development of the fungus during successive years, the shoots of the young Pines often become completely deformed; portions of them die off wherever the disease extends all round the shoot so as to stop the circulation of sap, and the young plants assume a somewhat frost-bitten appearance. The side-shoots which endeavour to assume the place of the leading shoots also usually become infected.

This fungus has an undoubted change of generation with *Melampsora tremulae* occurring on Aspen, whose teleutospores, produced in the following spring, develop the original form *Caeoma pinitorquum.*

Dry warm weather retards and checks the development of this disease; but when, for several seasons, the spring and early summer months are wet, it often occasions damage whose traces never become obliterated.

Prevention of this disease is confined solely to the selection of suitable soil and situation for the different species of Pine; whilst remedial measures include the removal and burning of all shoots infected, and the cutting out of all Aspen on the foliage of which the change of generation as *Melampsora tremulae* may be carried out (see also note to p. 282).

4. *Aecidium elatinum*, Link., the **Silver Fir Fungus**. To this fungus are due not only the peculiar cankerous, diseased swellings occurring on the stems at a greater or less height, but also the deformities or diseased twig-clusters, called '*witches-brooms*' in Germany, often formed in the crowns of Silver Firs.

The canker makes its appearance on the stem at all heights as well as on the branches. It is most frequent on loamy soils and low-lying localities, where the progress of the disease is more rapid than on sandy soils and upland tracts. It forms spindle-shaped excrescences—usually extending all round the stem, though sometimes only on one side,—on which the bark fissures longitudinally, so that the wood is laid bare and begins to rot. This process is hastened by insects becoming attracted, and by other fungoid diseases (*Polyporus*) effecting an entrance at the wounds; whilst the mycelium asserts itself throughout the bark-parenchym and the neighbouring sap-wood,—the same kind of mycelium as is found occasioning the twig-deformities. Infected stems become more or less unsuited for technical purposes according to the part attacked and the number of cankers, of which two or three may sometimes be seen on one tree; such stems also often break at the diseased parts during stormy winds. The canker may continue active in trees for fifty years or more, although during very hot summers it is apt to become fatal to the tree.

Another form of the disease occasioned by this fungus is

the occurrence of yellowish-green loranthus-like deformities
or twig-clusters of a more bushy appearance than the normal
branches from which they protrude at right angles ; the yellowish-green needles, growing closely and all round the twigs, are
shed during the first autumn. These deformities always have
their origin in a swelling or node within which the mycelium
of the fungus infests the bark, liber, and young wood, and
from which it annually extends to the young twigs and needles
till the disease dies out.

Whether a canker-spot or a twig-deformity will be produced
by the fungus depends on the nature of the place in which the
mycelium develops. Should the latter in extending itself effect
an entrance into a bud still capable of development, a deformed
twig-cluster is the result ; but when the shoots are already
formed, the mycelium is principally confined to the bark, and
a cankerous swelling is there produced. In either case infection seems only possible at wound-surfaces. The receptacles
(*spermogonia* and *stylospores*) develop in the diseased leaves of
the youngest shoots, appearing on the lower side of the needles
in June in the shape of red or orange cups (*aecidia*), which
burst and scatter their spores. All attempts hitherto to produce the disease directly by means of infection with these
spores have failed, as a heteroecious change of generation is
necessary ; but the intermediate form of the fungus and the
host on which it produces itself remain, for the present,
unknown to mycologists.

The formation of mixed forests is the best prophylactic
measure against this disease ; whilst the removal of twig-
clusters in June and July, and the cutting out of cankerous
stems, should be undertaken, wherever possible, in order to
check its spread.

5. *Nectria cucurbitula*, Fr., **the Spruce-bark Fungus**. This
is a cankerous disease occurring principally on the stems of
young Spruce from about 3 to 15 ft. in height, especially in
localities exposed to danger from frost. Infection can only

take place when the spores obtain a foothold on wound-surfaces occasioned by insects, hail, &c. ; hence the spread of the disease is greatest in places where the young crops are not vigorous in growth, where insects are likely to be attracted, and where any slight wounds inflicted take longest to heal.

The first symptoms of the disease are bleaching of the needles, and drying up of the bark and cambium which become brown, more particularly near wounds caused by the caterpillars of the Spruce-bark Tortrix (*Grapholitha pactolana*). This is followed by the appearance of numerous groups of small red concave *perithecia* on the bark which contain the tube-spores. Throughout the winter months these are scattered abroad, and spread the disease wherever they happen to land on open wound-surfaces.

The branching mycelium extends principally throughout the vessels of the new sap-wood and among the intercellular spaces. The whole development of the fungus takes place quickly, although it appears to be accomplished only during the period of rest, and not whilst the cambial layer is in full activity during the summer months. When the disease proceeds round the whole stem, the crown, and often the whole plant, dies off; but, if a strip of bark still remains sound, a cortaceous formation takes place at the edges. This prevents the further spread of the disease; and after the dead bark of the other part has been shed, the ordinary process of cicatrization begins.

The only means of combating this disease is to remove and burn, during autumn and the early months of winter, all shoots or young stems attacked. In doing this, care should be taken to remove them in the manner least likely to scatter the spores abroad.

6. *Peziza Willkommii*, R. Hartig, **the Canker of the Larch**. Although a diseased condition of the Larch is caused by bad attacks of the Larch aphis (*Chermes laricis*) or the mining moth (*Coleophora laricella*), yet the cankerous ailment occasioned primarily by this fungus is widespread and very serious in every

respect.   It occurs chiefly in damp, close localities, or on low-lying tracts, wherever mists and late frosts in spring are apt to prevail.   It is most frequent in young crops from ten to twenty years in age, though stems are still liable to become infected up to about forty years.   Hartig states[1] that the very great extent to which this disease occurs in the forests of northern Germany, Scotland, &c., is due to the more luxuriant development of the fungus, and to the less vigorous growth of the Larch ; for, though the *Peziza* is also to be found through-out the alpine home of the Larch, the fruits of the fungus generally dry up there without attaining maturity, whilst at the same time the tree is better able to resist the attacks of this parasite[2].

The first signs of the disease consist in the appearance of smooth, lustrous spots or slight swellings on the stem or branches, and the formation of longitudinal fissures in the bark from which a slight resinous outflow takes place.   This fissuring soon becomes more general, and bits of bark begin to scale off from the stem ; whilst small fungi appear from the fissures with cup-shaped apothecia having white felty edges and bright red centres.   The dead, diseased parts gradually grow scurfy and black ; whilst the wounds appear to deepen as the bark curls up at the edges of the fissures, and the canker gradually spreads up, and down, and round the stem.

It is only at damaged parts, such as wounds from insects (principally tree-lice and the mining moth), hail, bending of the branches under snow or ice, &c., that the spores can find a favourable germinating bed, from which the mycelium may penetrate into the cortex during the non-active period of vegetation.   Its advance is checked during the active time of the year by the formation of a tough corky layer around the diseased part ; but, during the following autumn, the mycelium again penetrates from the cambium into the bark, and enlarges

---

[1] *Lehrbuch der Baumkrankheiten*, 2nd edit. 1889, p. 45.
[2] See Chapter II, p. 43.

the canker. If the cankerous spot finally extend all round the stem the tree dies off above the diseased ring, though otherwise the struggle between the disease and the vitality of the stem may often be continued for decades. But, in large cankers, the mycelium is apt to penetrate through the medullary rays into the heart-wood of the stem, and to assume such proportions as to interfere considerably with the circulation of the sap, in which case the tree begins to sicken and die.

Dry situations and warm sunny exposures are unfavourable to the maturing of the sporophores; whilst humid localities favour the ripening of the spores. The disease is also of more frequent occurrence in pure crops of Larch, and whenever the canopy is close. The best preventive measures are therefore to be found (1) in careful choice of the soil and situation most suitable to Larch, and (2) in its growth in mixed forests only —along with broad-leaved trees, if possible, in preference to other conifers. Remedial measures are practically confined to the removal of all cankerous individual stems during the operations of thinning.

B. On Broad-leaved Trees :—

1. *Nectria ditissima*, Tul., **the Canker of the Beech.** Cankerous places on Beech-stems may be occasioned by tree-lice (*Lachnus* and *Chermes*), or by frost, or by sun-burn, or by fungoid disease; but in the latter case the disease is recognizable through the small white *Gonidia*-pustules which first of all appear, to be followed afterwards by numerous dark-red globular sporophores on the cankerous spots.

Although occurring also on Oak, Ash, Hornbeam, Alder, Lime, Maple, and Sycamore, as well as on Apple and Cherry trees in orchards, this disease chiefly attacks the Beech. It is not confined to any particular period of the development of the tree, being met with in mature crops, as well as in very young thickets, and sometimes on the one-year-old shoots, or sometimes on the bark of much older portions of the stem.

Infection appears invariably to take place at wound-surfaces occasioned by causes of the most various nature (hail, removal of standards, snow-pressure, injuries during clearing, weeding, thinning, &c.).   From such spots, frequently at the fork of young branches, the mycelium penetrates into the woody tissue and spreads quickly in a longitudinal direction up and down the stem, when the wood infected becomes brown and dead. As the mycelium develops, the canker seems to deepen into the wood; whilst the edges round the diseased part become raised, and a hypertrophic condition ensues, which results in spindle-shaped excrescences along the part attacked.   The branches or stems infected gradually become more and more cankerous, until finally the pole or young tree is killed off.

In order to prevent the spread of this disease, care should be exercised in the felling and removal of standards, or of parent trees above young seedling growth; whilst all cankerous Beech and Oak should be removed during the weedings and thinnings, so far as possible without interrupting the leaf-canopy.

Minor diseases are also occasioned by *Nectria cinnabarina*, a fungus very frequently occurring as small bright-red pustules on the twigs and buds of Maple, Sycamore, Elm, Lime, and in particular Horse Chestnut—also by species of *Polyporus*, the most important of which is *P. sulphureus* occurring on Oak, Poplar, Tree Willows, Birch, and Larch.   These species of *Polyporus*, which also occasion red and white rot in timber, are partly saprophytic, living only on dead wood, and partly true parasites causing disease.   In the latter case the mycelium develops in the interior of the stem; whilst the well-known fructification of different shapes, often bracket-like, are situated on the outside of the stem.   The timely removal of trees thus infected with fungus is advisable, not only in order to have any use from the timber itself, but also for the purpose of preventing the further spread of the disease through the formation and scattering of spores.

III. Fungoid Diseases of the Roots and the Base of the Trunk.

A. On Conifers :—

1. *Agaricus melleus*, L., the **Honey Fungus**, one of the edible Mushrooms. This is a very common and, in many localities, a very dangerous parasite in young coniferous crops, especially on Spruce, Scots and Weymouth Pines, and Larch, though seldom to be found on Black Pines. It also occurs extensively as a saprophyte on dead stools and roots of old trees, particularly on Beeches. It appears mostly in young crops of from four to fifteen years of age, but may also be found in mature stems of 100 years old. Close sowings and plantations formed with wisps of seedling Spruce suffer most from the disease. It is apt to be frequent in localities where, owing to deterioration of the soil, a coniferous crop is formed after a fall of broad-leaved timber, on the roots and stools of which the mushrooms or 'toad-stools' have developed saprophytically.

The first symptoms of the disease consist in the gradual yellowing, withering, and shedding of the needles. This is followed by withering of the shoots, swelling of the root-stool, fissuring of the bark, resinous outflow on the ground, rotting of the cambium, and finally the death of the plant, which usually occurs either between the middle of April and the middle of June, or else between the middle of October and the middle of November.

The blackish-brown mycelial filaménts of this partly saprophytic fungus, by extending themselves under the bark of the stool and roots, and throughout the soil, manage to effect a foothold on the roots with which they may come in contact; and then, boring their way through the bark, they commence their attacks as a parasitic fungus by forming long, white, flat rhizomorphous mycelial filaments developing between the sapwood and the cambium. Such rhizomorphous developments of the mycelium only occur in a limited number of fungi. In *Agaricus melleus* they assume the forms of *Rhizomorpha fragilis* var. *subterranea* so long as they are blackish-brown and lie outside the bark ; but they become *Rhiz. frag.* var. *subcorticalis* when the mycelial filaments develop below or within the bark. The sporophores of the former, appearing here and there above the soil, are less numerous, and less apt to attain maturity, than those of the latter variety, which make their appearance in thick clusters near the neck of the root. The blackish-brown filaments of the former are generally round, and ramify throughout the soil like fine rootlets ; whilst those of the latter are whitish and ribbon-like.

Concerning this fungus, Professor Vines makes the following remarks [1] :

' The most remarkable sclerotia are those of *Agaricus melleus*, a Fungus which is very destructive to timber. The mycelium gives rise to dark-coloured compact strands of hyphae, of the pseudo-parenchymatous structure characteristic of sclerotia; but they are peculiar in possessing continued special growth, and by this means they soon become long filaments, known as Rhizomorpha. It is in this way that the Fungus spreads from tree to tree : the Rhizomorpha-filaments grow under-ground from the roots of an infected tree to those of a healthy tree (usually a Conifer); it penetrates into them and spreads in the tissues external to the wood in the form of a white fan-shaped mycelium. The compound gonidiophores (*Agaricus melleus*) are borne either on the subterranean Rhizomorpha-filaments, or on the parasitic mycelium ; in either case the gonidiophores come to the surface.'

Young plants, when attacked, usually die off during the first year of the attack. A microscopic examination of them will then show that all the cambial layer and the resin-ducts have been invaded by the mycelium, robbed of their contents, and gutted to form empty hollows, the starchy matter in the cells being transformed into turpentine, which causes the resinous outflow on to the soil. The receptacles, mushrooms, or toad-stools (*pilei*), make their appearance in October, being developed

---

[1] *A Students' Text-book of Botany*, 1894, p. 314.

most numerously during damp seasons. They are of a dirty-brown or honey colour, with darker scales ; whilst the lamellae are yellowish-white, with reddish-brown spots forming later on.

Characteristic features of this disease are its occurrence here and there in patches, and the rapidity with which it kills young plants in full vigour, after they have made good growth in height during the season ; hence damage from this cause is easily distinguishable from death caused by insects, drought, or the like, when the plant only succumbs after a longer or shorter period of sickly growth. To prevent the occurrence of the disease, the stumps of broad-leaved species should, if possible, be carefully grubbed up, before sowing or planting coniferous crops.

The best practical means of preventing its spread consists in the pulling up of the plants attacked with all their roots, and burning them ; whilst the isolation of these places may be arranged for by digging small trenches round them to a depth of $1-1\frac{1}{2}$ ft. in order to hinder the subterranean extension of the mycelium. Careful collection of the older saprophytic mush-rooms on old stumps is also advisable, and none the less so on account of their being edible. Where blanks have been occasioned in young coniferous crops by this disease, it is best to fill them with transplants of broad-leaved species, as conifers are more exposed to a recurrence of the disorder.

2. *Trametes radiciperda*, R. Hartig, **the Root Fungus.** This is one of the most serious of fungoid diseases in coniferous woods. Scots and Weymouth Pines and Spruce suffer in a much greater degree from it than Silver Fir and other species. Young plants, poles, and trees, are seen to rot away suddenly near the roots and die off; whilst other individuals near them are also attacked and succumb, often causing serious inter-ruption of the canopy and sometimes blanks of considerable size.

Infection proceeds as a rule from the diseased roots of a neighbouring stem ; but it may also take place direct from *gonidia* or spores conveyed easily on the furry coats of mice, &c., or by wind. The colourless mycelium develops throughout the cambium as well as the woody tissue of the roots and the base of the stem. The walls of the cells are penetrated and destroyed by masses of mycelial filaments, so that the whole root-system seems at times to be in a rotten, damp condition. This rottenness spreads from the roots upwards into the stem—except in the case of the Scots Pine, whose morbid formation of resin is so great as to check the upward progress of the mycelium. The disease is also noticeable externally by the mycelium protruding from the fissures of the bark in the shape of small yellowish white pustules. The sporophores or spore-producers, of a yellowish-white or pure white colour, are mainly to be found on the roots or at the base of the stem in characteristic, large, flat, excrescences which sometimes assume a bracket-like shape; in the old days of flint and steel these fungi made excellent tinder.

Against direct infection by spores, preventive measures can hardly be taken. Hence protective and remedial measures are alike to be found in the grubbing up and removal of all diseased material before it has time to produce the conidia, and in the more general introduction of broad-leaved species of trees into the coniferous woods. Isolation of the patches infected by means of ditches leads to prolific development of sporophores on the roots cut through; this measure can, therefore, only be recommended where these latter can be duly removed and burned before there is any chance of their bursting and scattering fresh spores abroad.

B. On Broad-leaved Trees :—

1. *Rosellinia quercina*, R. Hartig, the **Oak-seedling Fungus.** This disease attacks the roots of young one- to three-year-old seedling Oaks, especially in nursery-beds and during damp years. It occasions fading and drying up of the plants. The roots appear to be woven round with fine filaments, in the

vicinity of which the bark-tissue turns brown; whilst small, round, dark-brown or blackish pustules, about the size of a pin-head, make their appearance here and there on the main root. The drying-up process begins at the terminal leaves and continues downwards. The small round pustules on the tap-root are fruits (*sclerotia*) from which numerous brown thread-like rhizomorphous filaments proceed, encircling the roots and extending throughout the soil. By means of these sclerotia, a perennial form of mycelium, the fungus can retain its vitality from year to year, thus withstanding the effects of dry summer periods which would be fatal to the ordinary mycelium, to the development of which damp warm weather is essential. As the tap-root is protected against infection by a corky epidermis, the mycelial filaments usually effect an entrance by means of the tender side-rootlets, whence they push into the main body of the seedling. The spread of the disease can, however, also take place through the gonidia developed by the mycelium above the soil, or by means of the spores of the black, spherical *perithecia* produced on the surface of the soil or in the stems of the diseased Oak-plants.

This disease is best combated by the removal of all diseased seedlings from the nursery-beds, and by isolating all parts infected by means of narrow trenches about 1 foot deep.

The **Changes in Generation** which not infrequently take place have already been alluded to briefly; the more important of them may be here named, although space prohibits any details being given concerning their life-history. They are specially characteristic of the rusts (*Uredineae*), all of which are parasitic.

In forming spore-producers portions of the mycelium of *Agaricus melleus* develop into *Rhizomorpha fragilis*, of which there are two varieties, viz. *subterranea*, where the mycelial strands remain under the ground, and *subcorticalis*, where they are under the bark of trees.

Aecidium pini becomes *Coleosporium senecio* on species of

groundsel (*Senecio*), the spores from which scatter in the following spring, spreading abroad the germs of the Pine needle-
rust and the Pine canker ; hence it is advisable to remove
groundsel before it flowers in April. *Aecidium abietinum* has its
Teleutospore form as *Chrysomyxa rhododendri* on rhododen-
drons, and *Aecidium columnare* its change of generation with
the *Calyptospora Goeppertiana* on cranberries. What form is
assumed by the aecidial spores of *Aecidium elatinum*, or on
what plant they occur, has not yet been discovered.

Caeoma pinitorquum becomes *Melampsora tremulae* on
Aspen foliage, hence Aspen should be cut out of Pine woods
to prevent the development of the Pine-shoot fungus. The
Larch blight, *Caeoma laricis*, has also its *Melampsora* form on
Aspen foliage.

M. Betulina on Birch probably stands in relation with some
form of *Caeoma*. *Melampsora Hartigii*, the Willow-rust, has
first its Uredo-form, *Uredo salicis*, and then its Teleuto-form,
Caeoma ribesii, on gooseberries and raspberries ; but this last
form is not absolutely necessary for its regeneration. *Melampsora salicina* on the Goat-willow similarly becomes *Caeoma
evonymi* on the Spindle-tree.

Prevention and Cure of Fungoid Diseases in Trees.

The great sylvicultural authority on this matter, Professor
R. Hartig, expresses the following opinions [1] :—

' *The best prophylactic measures* against the outbreak and spread of
epidemics is the formation of mixed woods. Subterraneous and super-
terrene infections are best obviated when each tree is, as it were, isolated
by neighbours of different species. *A change in the species of tree* may
even seem advisable, under certain circumstances, on soils which have
been taken possession of by root-parasites, or which contain Teleutospores retaining their vitality for many years. *Avoidance of the conveyance*
of fungus spores by men and animals is recommended, especially during
operations in young tree-forest. *The therapeutic measures adoptable* after
any outbreak of fungoid disease occasioned by root-parasites consist

---

[1] *Op. cit.*, pp 55, 56.

partly in the timely pulling or *digging up* of the sickly plants, and partly in isolation of the infested patch by means of *narrow trenches*. But the best and most important general measure to be recommended is the immediate *removal of all diseased plants* from the woods, so as to hinder further infection being spread by their spores. *Keeping the woods clean is an essential condition to the maintenance of their health.'*

Even a very superficial study of the character and life-history of the more important fungi causing diseases among our forest trees cannot fail to show the immense importance of the maintenance of close canopy, so as to hinder a rank growth of shrubs, berries, and weeds on the soil, which may tend to reproduce and spread heteroecious fungi. The necessity for keeping the woods clean and free from suppressed material in a more or less sickly condition will also be evident, as also for cutting out the soft-wooded trees like Aspen and Birch whenever it may be suspected that Pine or Larch trees in the vicinity are suffering from fungous disease. But even if considerations as to fungoid diseases alone are not sufficient to urge that woods should be kept clean and free from sickly growth, it should also be borne in mind how very much greater and more frequently disastrous are the dangers at the same time incurred from noxious insects.

CHAPTER XIV

THE PROTECTION OF WOODLANDS AGAINST INSECT ENEMIES

Sylvicultural Entomology, a branch of the Protection of Woodlands thoroughly taught at all the Universities and Academies in Germany which provide for instruction in Forest Science, relates not only to the injurious Forest Insects, but also to the useful species which prey upon these and tend to restrain their increase in large numbers likely to damage the growth of timber crops. Under noxious Forest Insects generally, all the species which may perhaps be found at any time on woodland trees are not included, but merely such as occur more or less frequently in large numbers, and do perceptible damage to the growth and development of the timber crops.

The nature of the injuries inflicted on woodland crops by insect enemies is manifold; and neither the seedling growth, nor the young pole-forest, nor the mature crop, is safe from injuries varying from slight attacks on foliage, bark, and roots, occasioning merely temporary disturbance, up to the total destruction of crops over enormous areas. Even when extensive devastations do not take place, both young crops and trees may be so seriously disturbed in normal growth, as to occasion loss of increment, and the formation of blanks necessitating outlay for re-filling and improving.

X 2

The Life-history of Forest Insects.

Insects (*Insecta*) are animals with jointed feet (*Arthropoda*), which have a body consisting of three main sections, six legs (three pairs), and usually also four or two wings (two pairs or one), and which pass through various stages of growth. Most insects undergo four stages of development or *Metamorphoses*, each distinctly distinguishable from those which precede and succeed it. These are in turn (1) *Ovum* or egg, (2) *Larva*, grub or caterpillar, (3) *Pupa* or chrysalis, and (4) *Imago* or mature insect. When all these separate stages are passed through an insect is said to have a *complete metamorphosis*; whilst the comparatively few instances (forming the orders *Hemiptera* and *Orthoptera*) in which a distinct pupal stage is not apparent, as the larva gradually becomes transformed into the imago, are termed an *incomplete metamorphosis*. The former are called *metabolian* (*insecta metabola*), and the latter *ametabolian* (*insecta ametabola*), whose pseudo-chrysalides are known as *nymphae*. The metabolic class comprises about 95 % of all insects, and the ametabolic only about 5 %. The *Ova* or eggs vary greatly in shape, size, and colour. Ovideposition may take place singly, or in clusters or nests on different parts of trees; they are sometimes protected by some sort of covering. The *Larva* often makes its appearance from the egg within the course of a few weeks; but in many other cases it hibernates within the shell, and only issues during the following spring. Different kinds of larvae have special names. Those of most beetles, partly six-footed, partly without feet, are called *larvae*, except in the case of cockchafers (*Melolontha*), where they are termed *grubs*; the sixteen-footed larvae of butterflies and moths —those of Spanners (*Geometridae*) have only ten, and a few Mining-moths (*Tineidae*) have none—are called *caterpillars*; the eighteen- to twenty-two-footed larvae of Sawflies (*Tenth-redinidae*) with their tail-like extremities are named *tailed cater-pillars*; and the feetless larvae of flies (*Diptera*), which show no complex structure of the head are called *maggots* or *strigs*. Several changes of skin take place in the larva before it enters into the pupal state of rest.   In the case of beetles (*Coleoptera*) the structure and different sections of the mature insect are externally distinguishable ; whilst in moths and butterflies (*Lepidoptera*) they are masked or indistinct. The pupa is some-times to be found lying unprotected on the soil, under moss, in fissures or under scales of the bark, and at other times it is covered by a woven *cocoon*, often of large size, as in the case of some of the *Bombycidae* ; whilst with flies (*Diptera*), in place of any cocoon, the last larval skin forms a protective case.   The stage of development which lasts longest is that in which the insect hibernates.   This is usually the larval stage, although many injurious beetles hibernate as imagines.   Except in genera which hibernate either as ovum or pupa, these two stages usually last only from two to four weeks.   Thus the devastating Spruce moth (*Liparis monacha*) appears after two to three weeks' rest ; whilst the destructive Pine moth (*Gastro-pacha pini*) remains as a chrysalis for six to eight months from autumn till spring, and the cockchafer (*Melolontha*) even so long as for four years.   According to the *Morphology*, or structure of the Imago or mature insect, it is classified among the beetles (*Coleoptera*), butterflies or moths (*Lepidoptera*), hymenopters (*Hymenoptera*), flies (*Diptera*), neuropters (*Neuroptera*), hemip-ters (*Hemiptera*), and orthopters (*Orthoptera*).

As soon as the imagines appear, they begin to swarm and reproduce themselves.   The male generally dies soon after having entered *in copulâ*, although this is not the case with beetles, which very often hibernate as imagines, nor with bees, which live for four to five years.

The complete cycle from ovum to ovum is termed the *Generation* of any insect ; this varies greatly in different genera and species.   Thus it is said to be *multiple or manifold* as in the case of plant-lice (*Aphides*), and ichneumon-flies (*Ichneumonidae*), of which several generations are produced within the year; *two-fold or double* in the case of bark-beetles (*Scolytidae*) and Saw-flies (*Tenthredinidae*), which produce two generations in each year; *single, simple, or annual* in the case of most butterflies and moths (*Lepidoptera*), which yearly produce one generation; *biennial or two-yearly* in wood-wasps (*Uroceridae*), the Pine Resin-gall Tortrix (*Tortrix resinella*), and many longicorn beetles (*Cerambycidae*); or finally *mult-annual or polyannual* in the case of the cockchafer (*Melolontha vulgaris*), which takes three, and sometimes four years to complete its generative cycle. Occasionally two generations take place in three years, as in the case of *Tomicus bidens*; but this is most probably only exceptional.

Metabolic insects feed merely as larvae and mature insects, and in some exceptional cases (e. g. Spanish Fly, *Lytta vesicatoria*, Pine weevil, *Hylobius abietis*, &c.) are only injurious to woodland growth as imagines; but with ametabolic insects the nympha also requires food-supplies. In both groups the destructive power of the larvae is often excessive, as numerous species daily consume many times their own weight of foliage.

Classification of Injurious Forest Insects.

The scientific or morphological classification of insects is of comparatively little use to the sylviculturist, for whom a biological method of grouping them together is in many ways preferable. Thus they may be classed with reference to the species of trees on which they feed, as Spruce insects, Pine insects, &c.; this is, however, open to the objection that many insects are not *monophagous* merely, but also *polyphagous*, or even *pantophagous*, devouring everything (like *Liparis monacha* when occurring in vast swarms). Or they may be classed with reference to the nature of the damage they do, which may be either *physiological*, interfering with the imbibition, transpiration, or assimilation of the food-supplies and nourishment, as in the case of bark-beetles and moths, or *technical*, interfering with the utilization of the timber for technical purposes, as in the case of wood-wasps (*Siricidae*), longicorns (*Cerambycidae*), and certain *Scolytidae*; but this would be met by the objection that in such cases the damage depends on too many different factors (species and age of tree, time of attacks, surrounding local circumstances, &c.) to be convenient. Or classification may be made with reference to the extent of the damage usually done; but this would be a very unsafe guide, for some of the most destructive species (e.g. *Liparis monacha*) are often for many years seldom to be seen, until at last circumstances favour their development in such enormous masses that they devastate woodlands over large tracts of country. Or they may be grouped with reference to the age of the crops generally attacked (seedling-growth, thickets, pole-forest, tree-forest); yet this would also be unsatisfactory, as many species are dangerous throughout all stages of the development of the crop. Finally, and preferably, they may be classified with reference to the portion of the tree on which they principally effect injuries. Adopting this last-named biological method of classification, and commencing with the roots, we may group the most dangerous species as follows :—

Root-destroyers, including the grubs of the mole-cricket (*Gryllotalpa vulgaris*) and the cockchafer (*Melolontha vulgaris*), and the caterpillars of the Seed Owlet-moth (*Agrotis*).

Wood-borers, comprising the larvae of wood-wasps (*Siricidae*), cervicorn beetles (*Cerambycidae*), and certain *Scolytidae* like *Tomicus lineatus*, *T. monographus*, and *T. dispar*, also the caterpillars of certain moths like *Cossus ligniperda* and *Zeuzera aesculi*.

Bark-beetles, including most bark-beetles (*Scolytidae*) i.e. most of the true bark-beetles (*Bostrychini*), and the cambial beetles (*Hylesinini*), and several weevils (*Curculionidae*), which, often in both active forms, either destroy the cambial layer of the bark and the sapwood (e. g. most *Bostrychini* and several

Curculionidae), or else hollow out the pith from young shoots (e. g. some *Hylesinini* and *Tortricidae*).

Leaf-destroyers, including the caterpillars of most moths (*Lepidoptera*) and Saw-flies (*Tenthredinidae*), also leaf-beetles (*Chrysomelidae*), cockchafers (*Melolontha*), and Spanish fly (*Lytta*).

In addition to these main groups, there are also **Bud-destroyers**, consisting of several weevils (*Curculionidae*) and some leaf-roller caterpillars (*Tortricidae*), **Seed-destroyers** like the acorn-borer (*Balaninus*) and the larvae of *Grapholitha strobilella* and species of *Anobium* in Spruce cones, and also **Producers of hypertrophic Deformities and various Malformations** on foliage, shoots and fruits by gall-wasps (*Cynipidae*), gall-midges (*Cecidomyidae*), plant-lice (*Aphidae*), which, though frequently more easily noticeable, are as a rule in reality of such comparatively trifling physiological and technical consequence that they may almost be left unnoticed.

Although a biological classification according to the ages of the crops attacked seems less preferable than that above sketched, it may nevertheless be remarked that most weevils or rostral-beetles (*Curculionidae*), some leaf-rollers (*Tortricidae*), and the grubs of the cockchafer (*Melolontha*) genus of the lamellicorn beetles (*Scarabidae*) usually attack seedling growth and young plantations; whilst crops that have outgrown the thicket stage and have developed into pole-forest or tree-forest are most exposed to attacks from the majority of species of moths (*Lepidoptera*), and then of the many bark-beetles (*Bostrychini*) and cambial-beetles (*Hylesinini*) which follow in their wake when once physiological disturbances have been occasioned and the young trees grow sickly in consequence.   Many moths, however, exhibit an unquestionable amount of choice between pole-forest and tree-forest in commencing to feed. Thus, for instance, the caterpillars of the Pine owlet-moth (*Trachea piniperda*) and the Pine Span-worm (*Fidonia piniaria*) are always to be found in pole-forest crops first of all, whence they migrate to older crops only when they have greatly increased in number ; whilst on the other hand the Spruce moth (*Liparis monacha*) and the Pine moth (*Gastropacha pini*) attack the older woods first, and then on rapidly increasing proceed to devastate the younger crops also.

And though the rough classification, **slightly injurious, noticeably injurious**, and **very injurious**, is extremely indefinite and elastic (for *Liparis monacha*, one of the most destructive of moths, is often for decades a comparatively rare insect), yet sylvicultural experience abroad has shown that the Spruce bark-beetle (*Tomicus typographus*), the large brown Pine weevil (*Hylobius abietis*), the cockchafer (*Melolontha vulgaris*), the Spruce moth (*Liparis monacha*), and the large Pine moth (*Gastropacha pini*) must certainly be classified as **very injurious** species; whilst it would not be at all incorrect to include in such a category also the Pine Beauty (*Trachea piniperda*), the Bordered White moth (*Fidonia piniaria*), the Pine Sawfly (*Lophyrus pini*), the small brown, white-spotted weevil (*Pissodes notatus*), and others which, under ordinary circumstances, are usually regarded as merely **noticeably injurious**. The debateable ground between this latter class and those that would have to be termed **slightly injurious** is also very vague and indefinite. It depends entirely on circumstances ; and whenever the only safe motto relative to injurious insects—**Principiis obsta**— is neglected, any comparatively rare or innocuous species may ultimately develop into a downright scourge, especially where large pure forests of Pine or Spruce are their feeding-grounds. The **monophagous** species then become **polyphagous**, and with still greater increase **pantophagous**, devouring everything within their reach.

The places in which the greatest number and variety of insects will usually be found are the warm plains and uplands, especially those with sandy soil. Southern exposures, frost-holes, and crops on inferior soil or of poor growth from any cause, are usually the places where the danger of a rapid increase in the number of noxious insects is chiefly to be feared; hence the additional importance of keeping such woods or portions of woods carefully thinned and tended so as to remove any sickly stems before they become breeding-places for the insects.

As a rule conifers are much more exposed to attacks from noxious insects than broad-leaved species of trees. And as they are at the same time comparatively deficient in recuperative power, owing to their not being able to set apart such good reserves of starchy and nitrogenous matters, the damage suffered by them is usually much greater. Spruce and Pine suffer far more from insects than any other species of forest trees; whilst among broad-leaved kinds Oak, Beech, Poplar, and Willow are more damaged than the Birch, Alder, Ash, Elm, Maple or Sycamore. But there can really be no comparison between the devastations sometimes occurring in coniferous woods and the attacks to which broad-leaved crops are exposed. Even after Oak standards have been defoliated in June by armies of caterpillars of the Processionary moth (*Cnethocampa processionea*), the trees, thanks to their nutrient reserves, flush into leaf again in July; whereas, when about four-fifths of the foliage of Scots Pine and Spruce have been devoured, the ultimate recovery of the former is extremely doubtful, and the death of the latter is certain owing to the much smaller reserve supplies it stores up.

Whatever the species may be which forms the crop, damage done to young growth is, for obvious physiological reasons, more serious than when it occurs on older poles and trees; whilst attacks made in spring are more injurious than those taking place in summer or near the close of the active period of vegetation, when the buds have been formed and the nutrient reserves secreted for the succeeding year's growth. The extent of the damage often also varies according to the parts injured. If the foliage alone be attacked the loss may perhaps be confined to the mere deficiency in increment; whereas when the roots and the cambium are extensively damaged, the injuries usually result in the wilting and death of the seedling, pole, or tree.

The Numerical Increase of Injurious Insects.

The reproductive capacity or prolificness of forest insects varies considerably according to the species. But, fortunately, the most prolific varieties are not among the most injurious kinds, as the average rate of reproduction of the latter does not exceed from 100 to 200 ova, according to Ratzeburg.

Hard winters are not particularly fatal to insects, and on the whole may be said to favour their increase, especially with regard to beetles protected by their tough outer casing of chitin; for severe cold works greater havoc among the insectivorous song-birds than amongst the insects. The most sensitive to damp cold weather and raw winds are the naked larvae which have no hairy or protective covering; these are often killed off in great numbers during the processes of changing their skins. Warm, dry, genial weather, ungrubbed stumps remaining in the ground over extensive falls of mature timber, sickly crops of seedlings, saplings, poles, or trees, and suppressed or unhealthy individuals with a weakly flow of sap— such as may be found in woods that have suffered from windfall, snow, or ice-accumulations—also want of clearing, thinning, and proper tending of the woodland crops, are all exceedingly apt to encourage the speedy numerical increase of noxious insects, which often multiply with exceeding rapidity in such breeding-centres. Bark-beetles (*Bostrychini*) and cambial-beetles (*Hylesinini*) select for ovi-deposition stems that have been thrown or broken by wind, or newly-felled trees, or those already brought into a more or less unhealthy condition by reason of attacks of moth-caterpillars on the foliage; whilst the large Pine weevil (*Hylobius abietis*) chooses for its feeding-ground, when available, crowded crops raised from seed or sickly plantations, and at the same time selects the stumps of newly-felled trees for its breeding-place. Moths also, when laying their eggs, frequently give the preference to crops of backward growth on inferior qualities of soil, which are thus all the more liable to succumb entirely to the ravages of the caterpillars during the following spring. Such favourable breeding- and feeding-places will usually be found to be the centres from which swarms of noxious insects spread to other woodland tracts. Hence the necessity for careful tending even in crops where such outlay does not seem likely to prove directly remunerative. Slovenly, negligent treatment of any one portion of an extensive wooded estate endangers the well-being of all the growing-stock of timber throughout the whole area under forest within a wide circuit.

All circumstances, therefore, tending to the formation of such breeding-centres are only too favourable to increase in the number of injurious insects ; whilst hot dry summers and long-continued drought, by reducing the woodland growth to a somewhat weakly condition, decidedly add to the prolificness of moths and saw-flies.

Amongst the influences inimical to the numerical increase of noxious insects may first of all be noted damp, rainy, cold, raw, ungenial weather, while caterpillars are changing their skins, and when beetles and moths are swarming and entering *in copulâ*, although the influence exerted by cold in winter is comparatively slight even in the case of species which hibernate in the larval or the chrysalid stage. But as, before any species can make its appearance in unusual and excessive numbers, the balance of nature must of necessity have been in some way or another disturbed (e. g. by bad breeding-years amongst insectivorous birds and insects which may happen to be its particular enemies) a natural readjustment would always of itself take place, although this might not happen until irreparable damage had been done to the woodland crops. Thus, experience in Germany has shown (1) that bad devasta-tions by moth-caterpillars as a rule last for three years, to which a previous and a succeeding year of minor damage may be added (without, however, counting the subsequent ravages of the *Scolytidae* and *Curculionidae*, that have also to be reckoned with owing to the sickly state in which the still existing crops are left), and (2) that then parasitic insects (*Ichneumonidae*), fungoid diseases, and insectivorous birds also increase numerically to such an extent as to decimate and finally exterminate the plague in a very short time. Under ordinary circumstances, however, and especially in mixed woods (as compared with pure or mixed crops of coniferous species only) with their larger numbers of insectivorous song-birds, the tendency to excessive reproduction and increase of noxious insects is held in check by the presence of their natural enemies. So far, therefore, as may be compatible with other and sometimes more important considerations, all these natural enemies of injurious insects should be preserved and their numerical increase encouraged. The chief of these are :—

A. Among Mammals.

Bats (species of *Vesperugo*), which consume large numbers of cockchafers and moths; moles (*Talpa europea*) that destroy grubs and mole-crickets (*Gryllotalpa vulgaris*); Shrew (*Sorex vulgaris*), Hedge-hog (*Erinaceus europaeus*), Squirrels (species of *Sciurus*), Weasel (*Putorius vulgaris*), Pole-cat (*P. foetidus*), Stoat (*P. ermineus*), Badger (*Meles taxus*), and Fox (*Vulpes vulgaris*), which devour large numbers of beetles and chrysalides, as has often been proved by the examination of their stomachs and excreta.

B. Among Birds.

The most generally useful are the Cuckoo (*Cuculus canoris*) —which is the only bird that eagerly devours the hairy caterpillars of species like the Pine moth (*Gastropacha pini*)—the Starling (*Sturnus vulgaris*), Flycatchers (*Muscicapa*), Tits or Titmice (*Paridae*), Tree-creepers (*Certhia familiaris*), Swallows (species of *Hirundo*), and most singing birds; then Thrushes (species of *Turdus*), Rooks (*Corvus frugilegus*), the Stannel-hawk (*Tinnunculus alaudarius*), and Wasp-buzzards (*Pernis apivorus*), Woodpeckers (species of *Picus*), Sparrows (*Passeres*) and Finches (*Fringillidae*), Crows or Ravens (*Corvus corax*), Jackdaws (*Monedula turrium*) and Larks (*Alaudidae*), which are only of minor utility in this manner.

C. Among Insects.

The class of **useful forest insects** is, on the whole, of consider-ably more importance in keeping a check on the abnormal increase of injurious kinds than either mammals or birds. They comprise not only the various *predatory species* which, often both as larva and imago, prey on the ova, larvae, pupae, and imagines of noxious insects, but also the *parasitic species*, whose eggs are generally deposited on the ova and larvae, or less frequently on the pupae, or imagines, of the injurious kinds, that thus become the *hosts*, on which the maggots prey when they issue from the ovum. By the careful observer these species will usually be found to exist in woodlands in much larger number and variety than might be imagined. Hence when anything like abnormal increase of the noxious kinds takes place, so also do these, their natural enemies, multiply in consequence of the greater amount of available nourishment and the more frequent opportunities of reproduction thus offered to them.

1. **The Predatory Species** include, among the *Coleoptera*, the sand-beetles (*Cicindelidae*), the predaceous land-beetles (*Cara-bidae*)— a large family among which many species of *Carabus* and the tree-climbing *Calosoma sycophanta* and *C. inquisitor* are particularly useful in killing the caterpillars of the Pine, Spruce, and Processionary moths—the dung-beetles (*Staphylin-idae*), the carrion-beetles (*Silphidae*), the hister-beetles (*Hister-idae*), the nitid or shining beetles (*Nitidulidae*), the thread-beetles (*Colydiidae*), the soft beetles (*Malacodermata*), the gold-beetles (*Cleridae*)—especially *Clerus formicarius*, which is so frequently

to be found in the larval galleries of the Pine-beetle (*Hylurgus piniperda*) under the bark of Scots Pine—and the lady-birds (*Coccinellidae*), of which *Coccinella septempunctata* and *C. bipunctata* are of most frequent occurrence, whilst several species of the genus *Halyzia* are only to be found in the woods, particularly in Spruce and Pine woods.

Among the *Hymenoptera*, the digging wasps (*Sphegidae*) stun caterpillars, beetles, plant-lice, &c., with their stings, drag them to their holes, and then utilize them for breeding places (so that they are both predatory and parasitic); the wasps (*Vespidae*), and in particular the hornet (*Vespa crabro*)—itself injurious to Ash—prey on moths and flies; whilst the ants (*Formicidae*), which live in colonies of thousands in their breeding-mounds, consume enormous numbers of larvae of various descriptions.

Among the *Diptera*, the predatory flies (*Asilidae*), which breed principally in sandy localities, attack and feed on many other kinds of insects; whilst the humming or chirping flies (*Syrphidae*) are more frequently to be found in orchards than in the woods.

Although the *Neuroptera* contribute the scorpion-fly (*Panorpa*), the camel-necked flies (*Rhaphidia*), the gold-eyed fly (*Hemerobia perla*), and the ant-lion (*Myrmeleon*), which are all useful in preying on noxious insects (the last-named unfortunately also diminishes the number of useful ants), whilst the *Hemiptera* furnish the scaly and other bugs (*Pentatomidae* and *Reduviidae*), and the *Orthoptera* produce the dragon-flies (*Libellulidae*), yet the predatory species among these three orders are comparatively inactive as compared with those in the first-named three.

2. **The Parasitic Species,** the chief insect enemies of the noxious varieties, are to be found among the ichneumon-flies (*Ichneumonidae*) of the *Hymenoptera*, and the parasitic-flies (*Tachininae*), a group of the *Muscidae* family of the *Diptera*. In fact, it is mainly to fungoid diseases and to the *Ich-*

neumonidae that the suppression of the occasional enormous plagues of devastating insects raised in the pure coniferous forests of Germany, Russia, and Austria is ascribable.

Of about 5,000 species of ichneumon-flies nearly 1,000 are parasitic on noxious insects[1]. The female is provided with a long ovi-depositor by means of which her eggs can be laid either singly in the case of large species, or in considerable numbers for smaller genera (290 larvae of *Microgaster globatus* have been counted in a single caterpillar of the Pine moth, *Gastropacha pini*). The eggs are more frequently deposited on the ova or the larvae than on the pupae or the mature insects. The tiny larvae subsist on the vital fluids of the hosts they are parasitic on, and then work their way out to the surface, where their cocoons may be seen thickly studding the dying caterpillars. As the generation of the *Ichneumonidae* is only partly simple, and in other cases manifold, they increase with enormous rapidity when the number of hosts is large. The larvae attacked become even more voracious than previously in order to maintain supplies of nourishment for their parasites, but perish during the pupal state when they succeed in passing through the larval stage of development.

The *Tachininae* are chiefly parasitic on the larvae and pupae of moths and saw-flies. The principal species are *Tachina monachae* on the caterpillars of the Spruce moth (*Liparis monacha*), and *Echinomyia fera* on those of the Spruce moth and the Pine Owlet or Pine Beauty (*Trachea piniperda*). They are easily distinguishable from other flies by the rough, brush-like hairs on their abdomen.

It would perhaps be asserting far too much to say that enormous swarms of insects might quite well be suppressed by *Ichneumonidae* and *Tachininae* unaided; for there can be little doubt that, in addition to killing larvae outright, the fungoid diseases predispose caterpillars to attacks from these parasitic

---

[1] Hess, *Der Forstschutz*, vol. i. 1887, p. 213.

insects. Experience of past calamities, collated with the recent investigations made in Bavaria owing to the devastations of the Spruce moth, have shown that the chief fungoid diseases, by means of which nature reasserts the normal balance obtaining before any plague begins, are produced by species of *Botrytis* (which occasions *Muscardine, B. bassiana,* in silkworms) *Micrococcus, Bacterium, Isaria, Cordiceps, Saccharomyces,* and *Torula.*

Preventive Measures.

As a numerical increase of injurious insects is most of all to be feared in crops of backward development or sickly growth, all prophylactic measures for warding off any predisposition to disease or to interference with the normal transpiratory and assimilative processes must tend to their repression, and obviate any inducement to abnormal reproduction favoured by increase in the quantity of material offered as suitable feeding-grounds and breeding-places. The careful clearing, weeding, thinning, and general tending of all kinds of woodland crops at the different stages of their development are therefore necessary in order that all suppressed, unnecessary, diseased, or damaged individuals may be eliminated before they offer attractions to insects which may ultimately be fatal to the more vigorous and healthy portions of the crop. Such measures of tending will, of course, be least needed wherever, at the time of the formation of the woods, due consideration has been given to the choice of the species of trees best suited for the varying conditions of soil and situation throughout the area to be brought under timber, and to the invaluable advantages to be gained by the formation of mixed woods[1] of broad-leaved species and conifers (wherever possible with any given conditions of soil and situation) in place of pure crops, or of mixed crops of conifers only. The crops should be inspected frequently;

[1] See lecture *On the Advantages of Mixed Woods over Pure Forests,* para. 7, p. 135.

whilst thinnings should take place as often as funds permit, and considerations relative to the maintenance of a full leaf-canopy and the safeguarding of the productive capacity of the soil dictate.   Wherever damage has been done by frost, snow, ice, wind, or fire, measures should not be delayed in rectifying it, so far as may be possible; and care should be taken to have the injured timber either barked, or, better still, removed from the vicinity of the woods as soon as convenient, in order that species of *Bostrychini* and *Hylesinini* may not become attracted to them as favourable breeding-places.   Felling of coniferous crops during the summer and careful removal of the bark from the stems are therefore to be recommended.   But when, from one reason or another, winter fellings have to be made, then some of the logs should be left here and there with the bark on, in order to act as decoy stems.   These should be peeled during the following May, and the bark burned so as to destroy the various broods of *Scolytidae* that may have been deposited in it.

Wherever any possible market for firewood makes it feasible to grub up the stumps of coniferous species of trees, without incurring much outlay, this should be done immediately after the fall of the timber, so as to diminish the number of natural breeding-places offered to the dangerous classes of weevils (*Curculionidae*) and cambial-beetles (*Hylesinini*).   Even casual examinations of palings formed of coniferous poles will show how many breeding-places *Bostrychini* and *Hylesinini* find in the bark and cambium.   Barking of the poles, therefore, before they get dry, would not only help to reduce the number of insects generated, but would also otherwise contribute to the durability of the wood, and to its protection against attacks from saprophytic fungi.

But, whilst mixture of different kinds of trees in the formation of the crops, and careful tending, are the principal means of preventing the abnormal increase of any kind or kinds of injurious insects, everything should at the same time be done to promote the increase of their natural enemies, most particularly amongst the class of insectivorous birds. These are, however, naturally of somewhat infrequent occurrence in dark, sombre coniferous woods as compared with broad-leaved and mixed crops. And amongst these insectivorous birds there is none which deserves so much encouragement as the Starling (*Sturnus vulgaris*), for whose convenience and protection nesting-boxes should be hung or nailed up in the woods, to protect their eggs and young against cuckoos, cats, and other enemies. Since the recent devastations of the Spruce moth in Bavaria (1890-1892) there have been many thousands of such starling-boxes distributed throughout the State Forests in order to promote the breeding of these most useful birds.

Exterminative Measures.

As previously stated, endeavours must be made to put an end to every abnormal increase in the appearance of any noxious insect, before it acquires the power of reproducing itself in countless swarms. This requires some knowledge of the appearance, habits, and life-history of injurious species generally. It also involves constant attention being paid to all areas where plantations or fellings have been recently made, or where damage has occurred from wind or snow, as well as to all sickly and backward crops, which might easily develop into extensive breeding-places. The frequent revision of Pine and Spruce woods is indeed an absolutely indispensable measure. The discovery of noxious insects may take place by many different kinds of observations. Bore-holes on the stems, bore-dust, or cobwebs, or drops of resin on the bark or at the foot of poles or trees, all indicate their presence in the stems of standing crops; whilst excreta or damaged foliage along paths and in the ruts of cart-tracks, and a gradual thinning of the foliage, indicate the presence of moth-caterpillars in the crowns of the trees. Even before these last signs become distinct, the presence of more insectivorous birds than usual conveys a plain hint to the thoughtful observer. Towards evening any abnormal swarming of moths serves to put one on guard; whilst threatened large attacks of the Pine moth (*Gastropacha pini*) may be foretold by the frequency with which the hibernating caterpillars may be found under the layer of moss covering the soil.

Whenever thus forewarned, or when called upon to annihilate swarms of any injurious species that may have acquired numerical strength enough to seriously threaten or interfere with the vigorous growth and development of timber crops, the exterminative methods adoptable vary according to the habits of the insect occasioning the trouble and anxiety. Some knowledge of its life-history is requisite in order that the easiest point of attack against it on a large scale may be fixed on. As, however, the methods usually adopted against beetles differ essentially from those in use against moths, these two great classes may be treated separately as forming two distinct groups comprising the chief species of destructive insects.

A. Annihilation of Beetles (Coleoptera).

So far as many of the bark-beetles (*Scolytidae*) and weevils (*Curculionidae*) are concerned, which chiefly breed in conifers, the felling and barking of stems attacked, and the burning of the bark whose cambial layer contains the ova, larvae, pupae, and often imagines, are the first steps to be taken. The foresters and woodmen should be carefully trained to find out such infected stems, and should also be taught the principle and the practice of felling a tree here and there so as to form a decoy-stem, and thus prevent the beetles selecting healthy stems for ovideposition in default of sickly ones. The careful revision and subsequent burning of the bark must of course be duly arranged for, and not on any pretence overlooked.

Such decoy-stems should be felled and laid in position early in the season, before the insects are likely to swarm, as many species naturally seek after diseased and recently felled trees before attempting to deposit their ova in healthy stems, whose strong resinous exudation to a considerable extent threatens the well-being of their brood. Stems that have thus been placed as decoys in winter or early spring must be barked and removed during May and June; and fresh decoy-stems must be made to take their place in summer, so as to catch the second brood of species having a double generation. For such decoys, dominated or suppressed, but still healthy, poles or trees should be selected, and not half-dry stems that are already nearly dead; otherwise these present no great attractions to the females in search of breeding-places. It is also well to support them upon rests so as to raise them off the ground and let the beetles and weevils have full access to the lower side, which remains fresh and sappy after the upper half is becoming dry. The branches (which may also be used for decoy purposes for many species) should be lopped off the stems in order to obviate the evaporation which would otherwise be continued through the foliage. Timber from the winter fall may thus be used for decoy purposes; but it must be removed and carefully barked not later than the middle of May. Its removal alone is not sufficient; for then the broods develop normally, and increase prolifically.

The decoy-stems should be examined from time to time to see if they have been successful in luring insects to deposit their ova there. This may be noted by small drops of resin or heaps of bore-dust near the punctures and bore-holes; but pieces of the bark should also be removed occasionally to see how far the young brood has developed. When the largest larvae appear about half-grown the proper time has come for the bark to be stripped and burned, as until then there would always be the danger of beetles, whose ovi-deposition was not yet completed, being able to continue their reproduction elsewhere.

Attempts to collect the larvae can only be made in the case

of cockchafer grubs in nurseries; whilst the collection of the mature insects is practically confined only to beetles of the larger kinds, like cockchafers and Spanish flies (*Lytta vesicatoria*), which may be shaken down from the crowns of poles or young trees. Smaller species of beetles may also be caught by means of decoy bundles of twigs or bark, which are burned when infested; whilst the large Pine weevil (*Hylobius abietis*) is entrapped in narrow, upright-walled ditches, and then killed by treading on them or pouring boiling water over them.

B. Annihilation of Moths (Lepidoptera).

True day-butterflies (*Papilionidae*) are practically innocuous to woodlands. The various injurious species, Sphyngides, Spinners, Spanners, Leaf-rollers, &c., are all moths or night-butterflies. It is chiefly during the larval stage that successful efforts can be made for the extermination of swarms of moths, although something can also be done to reduce the number of ova, pupae, and imagines.

Sometimes the caterpillars may be collected merely by hand (the workmen being provided with old gloves in the case of hairy caterpillars like those of the Pine moth). This is the case with those which hibernate under the moss, or which may be brought down to the ground by shaking the poles or knocking on the branches of trees with padded mallets or the flat heads of axes, or which may be caught in isolating ditches. By shaking and knocking, the caterpillars of the Pine Beauty (*Trachea piniperda*) and the Pine Span-worm (*Fidonia piniaria*) may be brought to the ground for collection; and the shaking or tapping is most effective during cool weather, or early in the morning, when their foothold on the foliage is looser than during warm sunshine. In the case of the Spruce moth (*Liparis monacha*) the newly developed larvae may easily be killed in large numbers, whilst they are still clustered together in schools of about ten to twelve, before wandering upwards to the crown

in search of food. The so-called nests of caterpillars of the
Lackey moth (*Gastropacha neustria*) may also easily be crushed
or burned, as well as the cocoon-like clusters containing those
of the Processionary moth (*Cnethocampa processionea*).

When the attacks are severe but confined to small areas,
the migration of the caterpillars to neighbouring crops may be
prevented by isolating the infested portions by means of narrow
ditches—a method which, of course, can only be efficacious
if the leaf-canopy overhead be also interrupted. The deepest
ditches are required for the large Pine moth, when they are
made nearly 2 feet deep; but, in general, a depth of 12 to
15 inches is sufficient, care being in every case taken to ensure
the walls being cut perpendicularly, whilst holes should be
made here and there (also with clean-cut, upright sides) along
the sole of the ditch to catch the caterpillars and lessen their
chance of escaping up the sides. It is also advisable to cut
several traverse ditches within the area thus isolated, in order
to make the work of extermination as complete as possible.

One of the great lessons taught by the recent ravages of the
Spruce moth in Bavaria—third in importance only (1) to the
great advantages derivable from the formation of mixed woods,
and (2) to the necessity for taking active steps against injurious
insects whenever there is the slightest sign of the balance of
nature having been disturbed in any such manner as to permit
of their abnormal frequency—is the efficacy of forming rings or
girdles of patent glue [1] or viscous tar around all stems in infested
tracts, for the purpose of hindering caterpillars from ascending
to their feeding-grounds on the foliage, or, if once up, from
descending the stems for the purpose of entering into the pupal
state of rest, or for any other purpose. It has thus become by
far the most efficacious method of combatting bad attacks
of dangerous enemies like the Pine and Spruce (Nun) moths.

---

[1] *Leim* (Ger.) means *glue*, and not *lime*, as was the incorrect translation
given in the Foreign Office report on *The Nun Moth* (*Liparis monacha*) in
Bavaria, published in 1891.

The caterpillars of the former are thus prevented from ascending the stems in spring after hibernating under the layer of moss on the ground ; whilst the same is also the case when the latter spin themselves down to the ground on cocoon-like threads (like some of the *Tortricidae*). This they practically do at least once before they lose the power of spinning their gossamer filaments or get too heavy to be supported and borne downwards by the latter; and these transformations occur simultaneously, though gradually, about the third time of changing their skins.

The essential requirement about good viscous tar is that it should remain sticky, and without either running or getting hard and dry from exposure to rain and sun. Its efficacy is not dependent on the mechanical action of its stickiness, but on some inherent smell of one or more of the ingredients that is exceedingly repulsive to the caterpillars till the ring becomes hard and dry. Even if individual caterpillars can be found to face and cross it, their feet and mandibles get clogged with the gluey composition, and they die of inability to procure food for themselves. It is therefore essential that the rings or narrow girdles should remain sticky until the larval state of development has been completed, that is to say, for at least six to eight weeks (although practically good patent tar remains active about three times as long) without the surface oxydizing or becoming covered with a skin. It is a patent composition, whose ingredients must, however, be a very open secret, as over a dozen firms advertise it throughout Germany for delivery at about 7*s.* 6*d.* per cwt. This quantity suffices for about 2½ acres on the average, although of course young crops require more per acre than older woods [1].

Previous to the application of the patent viscous tar the bark of the stem requires to be cleaned, or cleared of loose

[1] For the benefit of those who may wish to try the effects of this patent tar or 'Raupenleim,' the following addresses of manufacturers may be given :—L. Polborn, 1 Kohlenufer, Berlin S. ; Schindler and Mützel, Stettin ; and J. M. Witzemann, Stuttgart.

bark, so that the glue may bite well on to the stem evenly all round without so much being required as on rough-barked stems. This cleaning, or 'reddening' as it is called in the case of the Scots Pine, takes place with iron scrapers at about breast-height, where a girdle about 4 inches is cleared, care being taken not to injure the stem by too rough treatment. This preparation should take place in winter and early spring in all crops where inspection has shown that Pine moth caterpillars are hibernating under the moss, or that deposits of Spruce moth ova are noticeable on the stems. About the end of March or the beginning of April the formation of the rings of patent tar should take place before any mild spring weather has quickened the ova to life or the caterpillars to renewed activity. It may be applied with hard brushes; but the best results were attained in Bavaria by the use of small wooden spuds about 1 to $1\frac{1}{2}$ inches broad, the rings or bands being smoothed off with similar spuds hollowed to a depth of about $\frac{1}{8}$th of an inch. Wherever convenient, a thinning out of the crop should first take place so as to remove all individual poles or stems not actually necessary for the maintenance of the full leaf-canopy or of the productive capacity of the soil. This not only reduces the cost of the operation materially, but also removes the class of stems most likely to favour the increased propagation of the insects.

Even in the case of species whose caterpillars hibernate on the ground (e. g. Pine moth) the removal of the soil-covering of moss has been found to do more harm than good, as it rapidly exposes the soil to deteriorating influences.

The collection of the chrysalides by hand is only practicable when they are formed on shrubs or in fissures of the bark near the base of stems, and can hardly be carried out on any large scale. A much more efficacious method consists in the driving in of swine into the woods. They eagerly devour all smooth-skinned pupae like the Pine Beauty and the Pine Span-worm lying under the moss, although they will not touch hairy kinds like those of the Pine moth, owing to the irritant properties of the hairs.

It is only when deposited low down on the stem that the ova of species like the Spruce moth can practically be collected and destroyed ; but under any circumstances, and especially on the fissured bark of Pine trees, many eggs and clusters of ova are apt to be overlooked. Prof. Altum, a celebrated sylvicultural entomologist of Northern Germany, recommends a daub of viscous tar as efficacious in killing the clusters of ova of species like the Spruce and Gipsy moths (*Liparis monacha* and *L. dispar*) and the Beech-spinner (*Dasychira pudibunda*).

Attempts to collect and destroy the restless moths themselves can in no way be considered at all satisfactory. A good many experiments were made in Bavaria in 1891 with strong electric lights placed before powerful exhaustors, but without any practical results being attained with the Spruce moth.

As previously remarked, bad attacks of moths seldom last for longer than three years in their full destructive intensity. But within that time enormous damage may be committed, and vast sums may have to be incurred for fighting against the moths. The latter are also almost always followed by enormous numbers of bark-beetles, cambial-beetles, and weevils, whose prolificness is stimulated by the vast quantities of dead, felled, and sickly timber which litters the woods in quantities not removable all at once. These vast breeding-places, sooner or later, attract such large flights of insectivorous birds, that market-gardeners and orchard-owners in other localities begin to complain of insect pests owing to the departure of starlings, finches, &c. ; whilst at the same time the balance of nature re-asserts itself by epidemics of fungoid diseases and through any unusual prolificness among the *Ichneumonidae* and *Tachinidae*.

Too much stress cannot be laid on the importance of the early discovery of any, even apparently slight, increase in the usual number of injurious insects, and of the necessity that then exists for adopting exterminative measures. Enemies that might during the first spring be annihilated at a moderate outlay, may necessitate a vastly greater expenditure in the following and subsequent years, not to mention the monetary losses involved by having to fell and reproduce crops that represent considerable capital but have not yet attained their normal marketable or financial maturity. Thus, during 1892, no less than £75,000 was spent in the State Forests of Bavaria on ringing stems with patent viscous tar for fear of a recrudescence of the attacks of the Spruce moth, whose devastations during the previous three years had involved the loss of hundreds of thousands of pounds in the destruction of the timber crops. Large quantities of timber had to be thrown on markets that were soon glutted, and in which prices sank very low; and immature crops had to be felled and removed in order to try and hinder the otherwise inevitable appearance of enormous plagues of bark-beetles, cambial beetles, and weevils, that would have been bred within the sickly stems and the stumps in the ground.

It is true that, with our comparatively limited woodland areas, we are not exposed to anything like these extensive calamities in Britain. But it is worth while learning the lessons which such devastations teach ; and these are, *firstly*, that the formation of mixed woods should be favoured to as great an extent as convenient (especially in the case of conifers), and *secondly*, that dangers from insect enemies are minimized when the woods are kept properly tended throughout all the various stages of their development.